LECTURES ON THE PHENOMENA OF LIFE COMMON TO ANIMALS AND PLANTS

Publication Number 900

AMERICAN LECTURE SERIES

A Monograph in

The BANNERSTONE DIVISION *of*
AMERICAN LECTURE SERIES IN
THE HISTORY OF MEDICINE AND SCIENCE

WIKTOR W. NOWINSKI, M.D., Ph.D.
Research Professor of Biochemistry
Director, Cell Biology Unit
Department of Surgery
The University of Texas—Medical School
Galveston, Texas

LECTURES ON THE PHENOMENA OF LIFE COMMON TO ANIMALS AND PLANTS

By

CLAUDE BERNARD

Member of the Institut de France and the Académie de Médecine
Professor of Medicine at the Collège de France
Professor of General Physiology at the Muséum d'Histoire Naturelle, etc.

VOLUME I OF TRANSLATION BY

HEBBEL E. HOFF
ROGER GUILLEMIN
LUCIENNE GUILLEMIN

CHARLES C THOMAS · PUBLISHER
Springfield · Illinois · U.S.A.

Published and Distributed Throughout the World by

CHARLES C THOMAS • PUBLISHER

BANNERSTONE HOUSE

301-327 East Lawrence Avenue, Springfield, Illinois, U.S.A.

ISBN 0-398-02857-5

Library of Congress Catalog Card Number: 73-4297

With THOMAS BOOKS *careful attention is given to all details of manufacturing and design. It is the Publisher's desire to present books that are satisfactory as to their physical qualities and artistic possibilities and appropriate for their particular use.* THOMAS BOOKS *will be true to those laws of quality that assure a good name and good will.*

Printed in the United States of America

N-1

Library of Congress Cataloging in Publication Data

Bernard, Claude, 1813-1878.

Lectures on the phenomena of life common to animals and plants.

(American lecture series, publication no. 900. A monograph in the Bannerstone division of American lecture series in the history of medicine and science)

1. Physiology, Comparative. 2. Life (Biology) I. Title. [DNLM: 1. Physiology, Comparative. QT 4 B518L]

QH501.B5213 574.1 73-4297

ISBN 0-398-02857-5

COURS DE PHYSIOLOGIE GÉNÉRALE DU MUSÉUM D'HISTOIRE NATURELLE

LEÇONS

SUR LES

PHÉNOMÈNES DE LA VIE

COMMUNS

AUX ANIMAUX ET AUX VÉGÉTAUX

PAR

CLAUDE BERNARD

Membre de L'Institut de France et de L'Académie de médecine
Professeur de médecine au Collège de France
Professeur de physiologie générale au Museum d'histoire naturelle, etc.

AVEC UNE PLANCHE COLORIÉE ET 45 FIGURES INTERCALÉES DANS LE TEXTE

PARIS

LIBRAIRIE J.-B. BAILLIÈRE ET FILS
19, RUE HAUTEFEUILLE, 19

LONDRES
BAILLIÈRE, TINDALL AND COX

MADRID
C. BAILLY-BAILLIÈRE

1878

TRANSLATORS' INTRODUCTION

AT THE PINNACLE of his fame as a Physiologist, with the completion of the *Introduction to the Study of Experimental Medicine* and his *Report on the Progress and the Direction of General Physiology in France,* and with all the honors that now came to him—election to the French Academy, nomination to the French Senate, and seats on innumerable committees and councils—Claude Bernard began to give thought to the future. His physical powers sadly depleted by nearly five years of intermittent illness, his family life unhappily proceeding to an inevitable separation from his wife and two surviving daughters, thoroughly indoctrinated by their mother, but with the relaxed enjoyment of a new friendship with Madame Raffalovitch, Claude Bernard began to recast his life.

Writing in a note to the *Report,* he outlines these new objectives:

> All the discoveries and all the works I have published are often, I myself recognize, in the state of simple sketches or even at times inadequate outlines. I believe that they have nonetheless exerted a useful influence on the progress of physiology by stimulating new investigations by a great number of investigators. But I would wish it to be known that the obscurities, the imperfections, and the apparent inconsistencies that can be found in my various works are only the result of the lack of time, of difficulties in execution, and of a multiplicity of impediments that I have encountered in my scientific development. For many years I have been preoccupied with the idea of resuming my various studies and setting them forth in their entirety, so as to point up the general ideas they contain. I hope now that it will be possible to accomplish this second period of my scientific career.

So, indeed, it was arranged. Bernard gave up his Chair at the Sorbonne to Paul Bert, and assumed a new Chair of General Physiology at the Museum of Natural History; Pasteur seems to have been the prime mover in this arrangement which gave Bernard new laboratory space and a new platform from which to speak for General Physiology as he continued as before to employ his Chair at the Collège de France to foster Experimental Medicine. The Revue des Cours Scientifiques began to publish more or less verbatim reports of his lectures at the Collège de France and at the Museum; it is not always known how the original drafts were made, but certainly Bernard corrected the proofs on all of them. In 1872 and 1873 his courses at the Museum were devoted to "Lectures on the phenomena of life common to animals and plants."

These were, in fact, his basic statement of the philosophy, the problems, and the future of General Physiology, and in 1878, in the last days of his life, he began to rework these lectures in the form of two volumes, the first of which had progressed far enough for him to correct the proofs on his death-bed, the other published a year later after reworking by Dastre.

In 1966 the Librairie Philosophique J. Vrin in Paris published a facsimile of the second edition (1885) of the first volume, choosing it for typographic reasons. While the minor reworking of some paragraphs in the second edition and a different convention for handling notes have made no essential changes in the work, we have chosen for our part to use the first edition for the translation we present here because it had been in John Fulton's Library for over twenty-five years before it came to the National Library of Medicine, which supplied us with complete copies of both volumes and with a grant (LM-00453) which helped in part to defray the cost of the translation. Let these be the reasons for dedicating the translation to the Memory of John Farquhar Fulton, for his humanistic approach to science, and to the National Library of Medicine, where so many such precious volumes are so freely made available for the advancement of medicine.

HEBBEL E. HOFF, M.D., PH.D.
Benjamin F. Hambleton Professor
Chairman, Department of Physiology
Associate Dean for Faculty and Clinical Affairs
Baylor College of Medicine
Houston, Texas

ROGER GUILLEMIN, M.D., PH.D.
Adjunct Professor of Physiology
Baylor College of Medicine
Houston, Texas
Dean and Resident Fellow
The Salk Institute
La Jolla, California

LUCIENNE GUILLEMIN
Formerly Associate in Historical Research
Department of Physiology
Baylor College of Medicine
Houston, Texas

FOREWORD

IN BEGINNING TO publish the *Course in General Physiology* that he had given at the Muséum d'Histoire Naturelle, Claude Bernard had proposed to develop a series to parallel the *Course in Medicine* given at the Collège de France. In the one he strove to establish experimental medicine; in the other he laid the foundations of general physiology. This was to pursue, in another aspect, one single objective, the study of life.

Death did not permit Claude Bernard to complete his project; it took him by surprise on February 10, 1878, while, in full command of his subject, he was correcting the final proofs of the present volume.

He had already determined its title, *Lectures on the Phenomena of Life Common to Animals and Plants,* but in reality it was more than that; it was a *Program for General Physiology.*

In this volume Claude Bernard has summarized the whole of his doctrines, and it is the most complete and the most systematic work that he leaves to the scientific world.

He had himself determined the arrangement of the volumes that were ultimately to appear; he proposed to publish a volume on *Fermentations, Combustions,* and *Respiration,* a second on *Nutrition* and *Organic Synthesis,* a third on *Sensibility* and *Irritability,* and a final one on *Morphology.*

The material that he had prepared and which he proposed to coordinate and develop will not be entirely lost to science.

Mr. Dastre, Adjunct Professor of Physiology at the Faculté des Sciences, who followed the laboratory experiments of Claude Bernard for many years, and who was associated in his work, will collect the scattered fragments—and take care of their publication, as, moreover, he has done for the publication of the *Lectures on the Phenomena of Life.*

J.-B. BAILLIÈRE AND SONS

20 February 1878

ACADÉMIE DES SCIENCES

ORATION BY MR. VULPIAN

Member of the Académie des Sciences

At the funeral of

MR. CLAUDE BERNARD

16 February 1878

Gentlemen:

THE ACADÉMIE DES SCIENCES, so sorely tried no more than a few days ago by the death of two of its most celebrated members, Mr. Antoine César Becquerel and Mr. Victor Regnault, has again been cruelly afflicted. The most illustrious physiologist of our time, Mr. Claude Bernard, died last Sunday, February 10, 1878, at the age of sixty-four years.

The emotion this death has evoked in all ranks of society, the alacrity of the public authorities to render solemn homage to the memory of Claude Bernard, the unanimity with which this homage has been rendered, the concourse of a saddened multitude at this funeral, all attest to the greatness of the loss that we have just experienced.

The Académie des Sciences has designated me to address in its name a final farewell to Claude Bernard. This sad duty, which I had to accept, I can accomplish in a manner worthy of the learned body for which I speak, only after attempting to measure the depth of the void that has been created among us.

Claude Bernard, born at Saint Julien near Villefranche on July 12, 1813, came to Paris around 1834 to devote himself to the study of medicine and surgery, and having been appointed Interne des Hôpitaux in 1839, returned to the service to which he had already been attached as Externe, the service of Magendie at the Hôtel Dieu. It was while attending the lectures of this famous physiologist at the Collège de France that he discovered his true vocation.

In place of the didactic courses in physiology that he had attended up to then, he saw, at the College de France, a professor perform experiments in front of his audience, not only to confirm facts that had already been acquired, but also, and more often, to study problems that had remained without solution. Instead of narrative physiology, it was animated, living and speaking physiology. It was the experiment itself, forcefully seizing the attention of those who were present and impressing ineffaceable memories on their minds; it was, moreover, a series of discoveries, full of interest, taking place, so to speak, before the student's own eyes.

The effect of lectures such as these was decisive. Claude Bernard felt himself to be an experimenter. He entered the laboratory of Magendie as a voluntary assistant; from the second year of his internship he became his official preparator. From this time on, Claude Bernard devoted himself completely to research in physiology, except for a period of discouragement when it seemed that a scientific career would never become open to him, during which he returned to surgery.

A memoir published in 1843 under the title of *"Recherches Anatomiques et Physiologiques sur la Corde du Tympan"* and his inaugural thesis for the Doctorate in Medicine defended in 1843 and entitled, *"Du suc Gastrique et de son Rôle dans la Nutrition,"* are his first publications. Thereafter Claude Bernard worked without cease; discoveries followed upon discoveries, and celebrity was not slow in attaching itself to the name of such a physiologist. At first he was deputy to his master Magendie at the Collège de France. In 1854 he was appointed Professor at the Faculté des Sciences with a Chair of Physiology created for him; that same year he was appointed Member of the Académie des Sciences, filling the place that became vacant as a result of the death of the surgeon Roux. The next year he was called to replace Magendie in the Chair at the Collège de France. In 1868 he left the Faculté des Sciences to occupy the Chair of Flourens at the Muséum, and in the same year he also replaced Flourens at the Académie Française. Most of the foreign societies and academies hastened to admit him to their associate membership, and he was appointed Senator, Commander of the Légion d'Honneur, and member of various foreign orders, but I do not emphasize these extra-scientific titles; he was one of those who honored the honorary distinctions he consented to accept.

Having achieved the most enviable of positions, he continued to work with the same ardor as at the beginning, and each year reported the results of his indefatigable experimentation. A few months ago he read to the Académie des Sciences a series of memoirs of the greatest interest on glycogenesis in animals, and at the moment when illness unexpectedly attacked him, he was pursuing new studies. Thus he died, it can be said, in the fullness of scientific productivity, and in the midst of our sadness and our regrets we are obsessed by the painful thought that death probably destroyed important discoveries which he would not have delayed communicating to us.

This is not the place to recall all the works of Claude Bernard. I must limit myself to emphasizing his principal discoveries and indicating the influence that he exercised on physiology and on medicine.

Ranking first of his studies is the series of his admirable investigations on the formation of sugar in animals. These are studies that created a new era in science. Not only did they uncover a phenomenon absolutely unknown until then, the production of sugar by the liver in all animals, but more than that, they illuminated with a clear light the mechanism of the influence exerted by the nervous system on inner nutrition. They are, moreover, the starting point of a new theory of diabetes. From the time (1849) when Claude Bernard made his first communication to the Société de Biologie on the formation of sugar in the liver, until last year, when he gave us a lecture on new studies on glycogenesis, he never ceased to occupy himself with this great question; and it can be said that all we know about it of any importance, we owe entirely to him. After having found that the liver formed sugar at the expense of the blood that traverses it, whatever the diet of the animal, he showed that this sugar is the result of the metamorphosis of an amyloid substance, whose presence in the hepatic organ he was first to demonstrate, a substance produced by the cells of the liver itself, to which he gave the name of *glycogenic material.* He then showed that the quantity of sugar provided by the liver to the blood of the hepatic veins varies according to whether the animal is in a state of health or of disease. He discovered that the piqûre of a particular point in the medulla exerts such an influence on the formation of sugar by the liver that the blood, loaded with an excessive quantity of this material, lets it escape through the kidneys, and the animal becomes diabetic. This entirely unexpected discovery created profound astonishment in the scientific world, which soon gave place to admiration as the feat announced by the French physiologist was confirmed by all experimenters. In a series of investigations of prodigious sagacity, he showed the pathways by which the lesions of the medulla, whose effects he had just pointed out, act on hepatic glycogenesis. Never had a more penetrating gaze been directed into the depths of inner nutrition.

He went still further. As I indicated before, he himself derived conclusions from his discoveries which he applied to medicine. He created a new theory of diabetes. To him this disease is due essentially to a disturbance in the functions of the liver, to an exaggeration of the production of glycogenic material, and to a parallel overactivity of the conversion of this material to sugar. This disturbance most often has as its cause a change in the functioning of the central nervous system. This theory of Claude Bernard became the starting point for the most interesting pathological investigations; and today, after searching discussion, it seems to be on the point of triumphing over the resistance of its opponents.

Alongside this great work, and at least in the same rank, posterity will place the studies of Claude Bernard on the sympathetic and on the innervation of the vessels. Before these investigations practically nothing was known about the action of the nervous system on the production of animal heat.

In 1851 Claude Bernard published his first experiments on *"L'influence du Grand Sympathique sur la Sensibilité et la Calorification."* He showed that the section of the cervical fibers of the sympathetic chain on one side produces, along with a congestion of the corresponding half of the face, a considerable increase in the temperature of the same region.

Perhaps none of the works of Claude Bernard show more clearly the instinct of discovery, and the inventive sagacity with which he was so richly endowed. Had not numerous physiologists sectioned the cervical sympathetic chain since the time Pourfour du Petit showed that this operation produced a constriction of the pupil of the corresponding side? But none of them had noticed that this section also produces an elevation in the temperature in the parts innervated by the cut fibers. Claude Bernard was the first to unravel this most remarkable phenomenon. He also taught us that the nervous system has a powerful influence on the temperature of the various parts of the organism. At the same time he discovered the influence of this system upon the vessels.

In showing that the section of the cervical sympathetic chain provokes a congestion of all the parts to which the nerve fibers of this chain are distributed, he showed the way. A few months later, during which he succeeded in finding the true mechanism of this congestion, Brown-Séquard reached the same conclusion in America, and was first to publish that the congestion and the augmentation of heat, which result from this experiment, are due to a paralysis of the muscular tunic of the vessels. The existence of vasomotor nerves was unquestioned from then on. Pursuing, as he always did, the consequences of this discovery, Claude Bernard taught physiologists and physicians the physiological role assigned to these nerves and the importance of this role. The heart, the central organ of the circulation, ejects blood into the arteries, and this blood, ceaselessly propelled by new cardiac pulsations, returns to the heart by the veins. The movement of the blood would have the same characteristics in all the capillaries of the body if the vessels that carried it to these capillaries were everywhere inert. But this is not so. Thanks to the vasomotor nerves, the vessels provided with a muscular coat can constrict or relax; these modifications can be produced in one place and not in another; there can be congestion or anemia in one organ while the circulation undergoes no change in other parts. The face can redden or pale under the influence of the emotions, without the rest of the circulatory apparatus being notably affected; the mucous membranes of the stomach can be congested separately so to speak, during digestion, to provide for the requirements for the secretion of gastric juice, and then return to the normal state; the brain itself, in the moments of intellectual activity, can become the site of a more abundant irrigation with blood, without any notable disturbance taking place in the rest of the circulation. It can be the same for all organs. These are phenomena whose mechanism holds no further secrets since the work of Claude Bernard.

But this is not all: it was reserved for Claude Bernard to make another discovery relative to the physiology of the vasomotor nerves, a discovery which if not more important, was surely more unexpected than that about which I have just spoken.

The vasomotor nerves that modify the caliber of the vessels by producing a constriction of their contractile tunic or by ceasing to act on this tunic, are by no means the only ones that exert an influence on these channels. Claude Bernard found that there exist other nerves which, when they are subjected to a functional or experimental excitation, act also on the vessels, but then cause a dilatation therein. These are the vasodilator nerves, as they have been called in opposition to the nerves whose excitation provokes a vascular constriction, and which have been named vasoconstrictors.

It was while pursuing investigations of the greatest interest on the physiology of the salivary glands that Claude Bernard was led to this remarkable discovery. Like Ludwig, and without knowing of his work, Claude Bernard had determined that the electrization of the chorda tympani produced an increase in the secretion of the submaxillary gland; but he noticed what had escaped the physiologist of Leipzig, that this electrization produced at the same time a considerable dilatation of the vessels of the gland. These vasodilator nerves, truly *nerves of arrest,* have not yet been found except in a few places; perhaps, as Claude Bernard thought, they exist everywhere and play a considerable role in the state of health or disease.

Claude Bernard's studies on the salivary glands were fruitful for science; I will single out here, among the other facts that he had discovered in the course of these studies, only the reflex actions that take place in the submaxillary ganglion separated from the nerve centers in the spinal cord. Thus, for the first time, he provided the demonstration of the much debated physiological autonomy of the sympathetic nervous system.

Another gland, the pancreas, had also attracted his attention at the beginning of his career. At that time only the most incomplete ideas were available about the physoology of the pancreas; one of the most remarkable properties of the pancreatic juice had almost entirely evaded the investigations of the experimenters, I refer to its action on fatty materials. Claude Bernard showed that of all the fluids that enter into contact with the foods in the digestive canal, the pancreatic juice is the one that exerts the most powerful action on fatty materials, to emulsify them and to make them such that they can be absorbed.

In an entirely different kind of investigation, Claude Bernard, although indeed preceded by the celebrated physiologists Magendie and Flourens, was again in true initiator. I wish to speak of his beautiful studies on toxic and therapeutic substances. It is to him, in fact, that we owe the proper methods by which the physiological actions of these substances can be studied, and by the most brilliant of discoveries he has shown us all of the implications that can be drawn from these methods. In a series of decisive experiments, he showed us that curare abolishes voluntary motions by paralyzing the peripheral extremities of the motor nerves, while respecting the nerve centers, the muscles, and the sensory nerves. Moreover, he taught us that carbon monoxide kills vertebrate animals by asphyxia, by fixing itself in the red cells of the blood taking the place of the oxygen therein and rendering them incapable of any further absorption of this gas. Finally, to speak only of the principal facts, I ought to recall his memorable studies on the alkaloids of opium and on anesthetics.

I have sought to emphasize the most important discoveries of Claude Bernard; but how many other works should not be analyzed to recall all of the services that he has rendered to science! I shall limit myself to citing his investigations on the pneumogastric nerve, on the spinal nerve, on the trigeminal nerve on the common oculomotor nerve, on the chorda tympani, on the facial nerve; investigations in the course of which he invented new procedures for experimentation such as the evulsion of these nerves, the section of the chorda tympani within the tympanic case, procedures which today bear his name. Unfortunately I cannot also mention his studies on recurrent sensibility and on the conditions, so interesting from the point of view of general physiology, which cause this phenomenon to vary. I shall content myself, however, with enumerating his investigations on blood pressure, blood gases, the variations in the color of this fluid during the state of inertia or functional activity of the organs that it traverses (glands, muscles) ; on the variations in the temperature of the parts in the opposite states of rest and activity; on the difference in temperature between the blood of the right ventricle of the heart and the blood of the left ventricle in mammals; on the selective elimination by the glands of substances introduced into the body, or of those which accumulate in the blood under the influence of certain morbid states (sugar in diabetes, coloring material in the bile) ; on the special characteristics and the particular role of the saliva from each salivary gland; on the influence of nerve centers on the secretion of saliva; on the secretion and the action of gastric juice and the intestinal juice; on the modification of the secretions of the stomach and of the intestines after the ablation of the kidneys; on the albuminuria produced by lesions of the nervous system; on the composition of the urine of the foetus; on the electric phenomena that are manifest in nerves and muscles; on the comparison of the processes of inner nutrition in animals and plants, etc.

In a word, there is hardly any part of physiology on which Claude Bernard has not left his mark deeply imprinted by discoveries of the greatest interest.

Thus the influence of Claude Bernard on physiology has been immense. It can be said without exaggeration that for nearly thirty years most of the physiological investigations that have been published in the scientific world have been nothing but more or less direct developments or deductions from his own work. In this regard, he has truly been, in the full sense of the word, the master of almost all the physiologists of his time.

His influence on medicine has been no less great. Innumerable works on pathology have been inspired by his physiological studies. In this direction, too, he himself again showed the way. In his theory of diabetes and in his investigations on uremia, on congestion, on inflammation, on fever, he indicated how the progress of physiology can serve that of medicine. His works have in many ways genuinely transformed the scientific part of medicine; his name is invoked in the history of a great number of diseases by the theories that have as their purpose either to explain the mode of action of morbid causes or to find the physiological explanation for symptoms. Therapy itself has felt the influence of his work. Nearly all medications have been subjected to new studies based upon his own investigations, and therapeutics has finally been able to merit the title of rational, to which until then it had no right. Such services cannot fail to be recognized. Thus medicine, which has always considered Claude Bernard as one of its own, as one of its most brilliant lights, regards his death as the greatest sorrow that could afflict it.

Shall I speak of Claude Bernard's publications, of the books in which his lectures at the Collège de France and the Muséum d'Histoire Naturelle are reproduced, or of his *Rapport sur les Progrès de la Physiologie en France,* published in 1867, on the occasion of the Exposition Universelle? What could I say that you would not all know? These books are in the hands of all physiologists and all physicians. They are perfect examples of their kind. Besides the original discoveries whose detailed accounts they contain, there are found therein, practically on every page, ingenious ideas, novel views, and important applications. There one participates in the evolution of the investigations of the master, from their first germ to their complete development, and while acquiring thereby a taste for personal investigation, one learns there how to work by himself.

Finally, after having spoken of the illustrious savant, should I not speak a word of the man? Is this not a duty, and the most pleasurable of duties, to recall that this physiologist of genius was at the same time the best of men? The simplicity of his manners, his affability, the steadfastness of his relationships, everything attracted one to him and made him loved. Free of any vanity, more than anyone, he knew how to render justice to the merit of others, and he was always ready to hold out his hand to young scientists to help them to climb the difficult steps that lead to official positions.

Such are the titles that Claude Bernard holds to the admiration of the scientific world and the gratitude of the nation. Posterity will place him among the number of the great men to whom physiology owes its most considerable progress, and his name will shine out beside those of Harvey, Haller, Lavoisier, Bichat, Charles Bell, Flourens, and Magendie.

In the name of the Académie des Sciences, dear and illustrious master, I bid you adieu!

FACULTÉ DES SCIENCES OF PARIS

ORATION BY MR. PAUL BERT

Professor at the Faculté des Sciences

At the funeral of

MR. CLAUDE BERNARD

16 February 1878

THE FACULTÉ DES SCIENCES of Paris, which for fourteen years had the honor of counting Claude Bernard among its professors, cannot remain silent beside this tomb, even though this illustrious master has been absent from its midst for ten years. It comes in its turn to express its regrets and claim its rightful share of glory.

It was in 1954 that Claude Bernard became one of us. The great discovery of the production of sugar by living beings had just struck the scientific world with surprise and admiration. To permit its author to develop all the resources of his fertile genius, a chair was then created which under the title of General Physiology came to increase and complete the scope of instruction in our faculty.

The valiant fighter had nevertheless obtained only a part of the requirements for free investigation. No material means for operation were attached to the chair in which he was to teach: neither budget, laboratory, nor assistant. Yet it was in the midst of this penury, an indictment of the indifference of the public authorities that from 1854 to 1868 Claude Bernard had to give his courses. He succeeded only by utilizing the resources of the chair he soon received at the Collège de France as an inheritance from Magendie.

Our faculty cannot therefore claim the honor of having seen the blossoming of those discoveries whose increasing accumulation rapidly carried his scientific reputation to the highest level. It was from the laboratory at the Collège de France, quite poor itself, that there emerged those innumerable works, any one of which would have been enough to make its author famous.

But if it was at the Collège de France that the creative genius of Claude Bernard was displayed in the domain of experimental research, it was manifest with no less force and utility for the general development of science through his teaching at the Sorbonne.

The foundation within this faculty of a chair of general physiology had given this experimental science its rightful place in classical teaching, at the side of its older sisters, physics and chemistry. It was to justify this new establishment, which had not been universally approved, that Claude Bernard devoted his lectures.

Before him, physiology had hardly been considered as anything more than an appendage of other sciences, and its study seemed to be the prerogative, according to the details of the problem, of physicians or zoologists. Some declared that anatomical knowledge of the organs was enough to permit deduction of the mechanism of their function, that is to say their physiology; others saw therein only an ensemble of dissertations, adequate to satisfy the systematic spirit about the causes, the nature, and the seat of various diseases. Nearly everyone attached to its teaching only a value, varying from one living species to another or within the same species, according to indeterminable circumstances; a value subject to the caprices of a mysterious and uncontrollable power, in this way denying, in reality, the claim of physiology to be a science.

Claude Bernard began by restoring this title to physiology. Most often using his own discoveries as examples, he showed that, although the questions it raises are more complex than in other experimental sciences, it is as sure of itself as they are. When, once the problem is posed, its elements collected, and its variables eliminated, it experiments, reasons, and concludes.

He showed that from the infinite variety of functional phenomena in relation to the diversity without number of organic forms, there emerge fundamental and universal truths which unite everything that lives in a common group, without distinction of orders or classes, of animal life or plant life; the liver making sugar like fruit, and brewers yeast going to sleep like man under the influence of the vapors of ether.

He showed that even for the physiology of mechanisms, anatomical deduction is insufficient and often misleading, and that only experimentation can lead to certainty.

He showed that the rules of such experimentation are the same in the sciences of life as in those of inanimate objects, and that "there are not two contradictory natures giving rise to two orders of science opposed to each other."

He showed that the experimental physiologist not only analyzes and demonstrates, but dominates and directs, and that he can hope to become, just as the physicist or the chemist, a conqueror of nature.

He showed that, although the physiologist must have constant recourse to the ideas provided by anatomy, histology, medicine, natural history, chemistry, and physics, he must remain their master, and subordinate them to his own views; thus he has need of a special education, of special instruments for research, of special chairs, and of special laboratories.

It is in this way that Claude Bernard gave a firm foundation to physiology, delimited its domain, banished capricious entities, rid it of empiricism, determined its goals, formulated its methods, perfected its procedures, and pointed out its means of action. He assigned to it its rank among the experimental sciences, demanded for it its legitimate place in public instruction; and in a word, he put it in possession of itself, gave it an individual identity, and characterized it as a science, living with it and identified with it to such a degree that a foreign scientist could say "Claude Bernard is not only a physiologist, he is Physiology."

Such is the part of the work of the illustrious physiologist, and it is not a small one, that our faculty can claim with pride. Such was, in fact, the substance of the teaching as he presented it until 1868, when he left the Sorbonne for the Muséum d'Histoire Naturelle.

It is to that one of his pupils who was called to succeed him in the Chair of Physiology that the faculty has confided the honor of representing it today. May he be permitted now to divest himself of his official role, and in the name of Claude Bernard's pupils, express a filial farewell to the master who is no longer with us. Indeed, he who owes the most, since he owes him everything, might almost demand this painful privilege as his right.

Certainly, Science and the Nation have reason to mourn. But what a deep sadness is added to these universal feelings, in the heart of those who have profited from his teaching, received the marks of his generosity, and experienced the effects of his paternal protection. Benevolent and sympathetic to all; he was, to those whom on his death bed he called his scientific family, the most affectionate and the most devoted of masters; not from an affection without restraint, because, generous in advice and encouragement, he showed himself as critical of our work as he was of his own; not from a devotion without sacrifice, because he suffered by his voluntary resignation from the chair at the Sorbonne, in order to leave it to one of his pupils. Among the daily incidents of the laboratory there was never an impatient word, never a bitter word in the face of so much physical and mental pain so courageously borne, never a reproach to those whose gratitude faded too soon! To his last days, in his last words, and in the face of that unexpected death, there were affection, counsel, and smiles; he thanked us for our care, we who owed him a hundred times as much! Continue to work, he said, and he spoke of the science that was his life.

Yes, master, we will keep on working; in the midst of our sorrow we all feel that growing duty. We will close our ranks. We will march forward, following your luminous pathway along the unfinished furrow.

CONTENTS

Page

Translator's Introduction vii

Foreword xi

Oration by Mr. Vulpian, Member of the Académie des Sciences at the funeral of Mr. Claude Bernard xxiii

Oration by Mr. Paul Bert, Professor at the Faculté des Sciences at the funeral of Mr. Claude Bernard xiii

COURSE IN GENERAL PHYSIOLOGY
OPENING LECTURE

Inauguration of general physiology at the Muséum—Reasons for the transfer of my chair from the Sorbonne to the Jardin des Plantes—Today, Physiology is becoming an independent science, separate from Anatomy—It is an experimental science—Definition of the domain of general physiology—Beginnings in France—Development of Physiology in neighboring countries—Laboratory installations—Necessity of a good method and a healthy experimental *critique* 3

LECTURES ON THE PHENOMENA OF LIFE
Common to Animals and Plants

FIRST LECTURE

I. Definitions in the sciences; Pascal. Definitions of Life: Aristotle, Kant, Lordat, Ehrard, Richerand, Tréviranus, Herbert Spencer, Bichat. *Life* and *Death* are the two states that are understood only by their opposition—The definition in the *Encyclopedia*—Life can be characterized but not defined—General characteristics of life: organization, reproduction, nutrition, development, aging, disease, death—Attempts at

Page

definition derived from these characteristics—Dugès, Béclard, Dezeimeris, Lamarck, Rostan, de Blainville, Cuvier, Flourens, Tiedemann—The essential characteristic of life is *organic creation.*

II. Hypothesis of life: spiritual and materialistic hypotheses; Pythagorus, Plato, Aristotle, Hippocrates, Paracelsus, van Helmont, Stahl; Democritus, Epicurus, Descartes, Leibnitz—The School of Montpellier—Bichat, etc.—We do not accept either materialistic or spiritualistic hypotheses in physiology because they are inadequate and alien to experimental science—Observation and experiment teach us that the manifestations of life are neither the product of matter nor of an independent force; that they result from the natural interaction of preestablished organic conditions, and precise physico-chemical conditions—We can apprehend and understand only the material conditions of this interaction, that is to say the *determinism* of the manifestations of life—Physiological determinism encompasses the problem of the science of life; it will permit us to master the phenomena of life, as we control the phenomena of inanimate objects whose conditions are known to us.

III. Determinism in Physiology—It is absolute in Physiology as in all experimental sciences—It was wrongly attempted to exclude determinism from the science of life—Distinction between philosophical and physiological determinism—Answer to the philosophical objections; physiological determinism is an indispensable condition for moral freedom instead of being its negation—Necessary separation of physiological questions from philosophical or theological questions—There is no possible conciliation between these separate problems; they are derived from different intellectual needs and are resolved by opposing methods—Neither the one nor the other can profit by a rapprochement 16

Page

SECOND LECTURE

The Three Forms of Life

Life cannot be explained by an internal principle of action; it is the result of a conflict between the organism and the ambient physicochemical conditions. This conflict is by no means a discord but a harmony—Life presents three aspects to us which prove the necessity for physicochemical conditions for the manifestation of life—These three stages of life are: 1) Life in a state of non-manifestation or latency; 2) Life in the state of variable and dependent manifestation; 3) Life in the state of free and independent manifestation.

I. *Latent life*—The organism falls into a state of chemical indifference—Examples taken from the plant kingdom and from the animal kingdom—Latent life is an arrested and not a diminished life—Conditions for the return of latent life to manifest life—Extrinsic conditions: water, air (oxygen), heat; intrinsic: reserves of nutrient materials—Experiments on the influence of air (oxygen)—Experiments on the influence of heat—Experiments on the influence of water—Phenomena of latent life in animals: infusoria, keronas, colpodas, tardigrades, anguillules of rusty wheat—Identification of the seed and the egg is not exact from the standpoint of latent life—Existence of beings in the state of latent life: brewer's yeast, anguillules, tardigrades, etc.—Explanation of the return from latent life to manifest life—Experiments by Chevreul on the desiccation of tissues—Mechanism of the transition to latent life—Mechanism of the return to manifest life—Necessary succession of phenomena of organic destruction and creation.

II. *Oscillating life*—Belongs to all plants and to a great number of animals—The egg shows dormant life—Mechanism of vital dormancy—Influence of the external environment on the internal environment—Diminution in the chemical phenomena

Page

during dormant life—Mechanism of vital oscillation in dormancy—Necessity of reserves for dormant life—Mechanism of vital oscillation—Cessation of dormant life—Influence of heat; it can bring on dormancy like cold—Resistance of dormant beings—Animals awakened during dormancy use up their reserves rapidly and die—Phenomena of creation and destruction in dormancy—Temporary dormancy does not require reserves like prolonged dormancy—

III. *Constant or free life*—It is the result of an organic perfection —Our distinction between *internal environment* and *external environment*—Independence of the two environments in animals with constant life—The improvement of the organism in animals with constant life consists in maintaining the intrinsic and extrinsic conditions necessary for the life of the elements within the internal environment—Water—Animal heat—Respiration—Oxygen—Reserves for nutrition—It is the nervous system that is responsible for the equilibration of all the conditions of the internal environment—Conclusions relating to the interpretation of the three forms of life—One cannot find a vital, independent force or principle. There is only one vital conflict, whose conditions we must seek to understand 46

THIRD LECTURE

Classification of the Phenomena of Iife

I. Classification of the phenomena of life—Two great groups: *organic destruction* and *creation*—This classification characterizes general physiology and embraces within its confines all the manifestations of life—Unity of life in the two kingdoms.

II. Classifications of the living beings; Linneus, Lamarck, de Blainville—Theories of the duality of life in the two kingdoms—Differentiation of the natural kingdom—Contrast be-

Page

tween animals and plants—Chemical, physical, and mechanical anthesis between animals and plants—Priestly, Saussure, Dumas, and Boussingault, Huxley, Tyndall.

III. General refutation of dualistic theories of life in animals and plants—Latest form of the theory of the duality of life—The quality of life and general physiology—Unity of the laws of life; variety of the manifestations of life, and the different functioning of living machines—Conclusion: the integrity of the phenomena of organic destruction and creation proves the unity of life 92

FOURTH LECTURE

Phenomena of Organic Destruction

Fermentation—Combustion—Putrefaction

Phenomena of *organic creation* and *destruction*—Study of the phenomena of organic destruction—Fermentation, combustion, putrefaction.

I. *Fermentation*—Catalysis; Berzélius—Decomposition; Liebig—Organic theory; Cagniard de Latour, Turpin, Pasteur—Soluble ferments, organized ferments—Actions of soluble ferments found in the mineral kingdom—The same ferments are common to the two kingdoms, animal and plant—The ferments act to transform and decompose the products of the nutritional reserves—Fermentation due to organized ferments—Alcoholic fermentation; its conditions.

II. *Combustion*—Theory of Lavoisier; direct, active or latent combustion—Direct combustion does not exist—Indirect combustion, cleavage, kind of fermentation appertaining to plants and animals—Particular case of glands—Unknown role of oxygen in the organism.

III. *Putrefaction*—Appertains to animals and to plants—Theories

Page

of putrefaction; Gay-Lussac, Appert, Schwann, Pasteur—Putrid fermentation—Analogy between putrefaction and fermentation—Life is a putrefaction—Mitscherlich, Hoppe-Seyler, Schutzenberger, etc. 113

FIFTH LECTURE

Phenomena of Organic Creation

Theories Anatomical: Cellular, Protoplasmic, Plastidular

Organic creation, comprising two orders of phenomena common to the two realms: *chemical synthesis, morphological synthesis.*

I. Anatomical constitution and morphological creation of the living being, animal or plant; history—Early period: Galen, Morgagni, Fallopius, Pinel, Bichat, Mayer—Modern period: de Mirbel, R. Brown, Schleiden, Schwann—Cell theory—The ultimate morphological element of living beings is the cell, but a living substance antedates the cell; this is the *protoplasm*—It is the seat of chemical syntheses, of morphological syntheses.

II. Origin of the cell in protoplasm—Protoplasmic theory—Blastema—Gymnocytode, Lepocytode—Protoplasm in plant cells—The primordial utricle—The protoplasm is the living body of the cell in the two kingdoms.

III. *Protoplasm;* its constitution—Protoplasmic mass, nucleus—Protoplasmic beings—Monera, Bathybius—Structure of the protoplasm—Plastidular Theory—Complexity of protoplasm—Its role in the division of the nucleus—Relationships of nucleus and protoplasm—The nucleolus, its constitution, its role—Conclusion 129

Page

SIXTH LECTURE

Chemical Theories—Syntheses

Colorless Protoplasm and Green, or Chlorophyllian, Protoplasm

Protoplasm and organic creation—Generalities—Chemicophysiological synthesis—Elementary constitution of organized bodies—Creative synthesis is necessarily chemical, but it has special processes—Green or chlorophyllian protoplasm and colorless protoplasm—They cannot serve to delimit the animal kingdom from the plant kingdom.

I. Role of chlorophyllian protoplasm in organic synthesis—It produces the synthesis of ternary bodies under the influence of light—The experiment of Priestley is the starting point of this theory—Hypotheses of the chemists on the subject of syntheses in green protoplasm—Green protoplasm derives its energy from solar radiation.

II. Role of colorless protoplasm in organic syntheses—It produces complex syntheses—Pasteur's experiments—It cannot incorporate carbon directly, however—Colorless protoplasm uses caloric energy—Status of the question of organic syntheses; new hypotheses—Hypothesis of cyanogen—Chemical synthesis and vital force.

III. Synthesis in particular—The best known example is amylaceous or glycogenic synthesis—Discovery of animal glycogenesis—Phenomena of synthesis of starch and destruction of starch—Principal characteristics of glycogenic synthesis in animals and plants 145

SEVENTH LECTURE

Properties of Protoplasm in the Two Realms

Irritability, Sensibility

Protoplasm possesses irritability and motility—These properties

Page

constitute the connection between the organism and the external world.

I. *History of irritability*—Glisson, Barthez, Bordeau, Haller, Broussais, Virchow—Irritability; autonomy of the tissues—Protoplasm is the seat of irritability.

II. *Excitants and anesthetics of irritability*—Normal conditions for protoplasmic irritability—Anesthesia (1) of protoplasmic properties of the activity of irritability or sensibility in animals and plants—Experiments—Anesthesia of the protoplasmic phenomena of germination, development, and fermentation in animals and plants—Anesthesia of the germination of seeds—Anesthesia of eggs—Anesthesia of organized ferments—The impossibility of anesthetizing soluble ferments—Anesthesia of the chlorophyllian function in plants—Anesthesia of the worms of rusty wheat.

III. *Irritability and sensibility*—Conscious and unconscious sensibility—Different points of view of philosophers and physiologists on this subject—Identity of the anesthetic agents abolishing sensibility and irritability—We do not act on the properties or functions of nerves, but solely on the protoplasm 173

EIGHTH LECTURE

Organized Synthesis, Morphology

Protoplasm represents life alone without specific form—Form is necessary to characterize the living being—Morphology is distinct from the chemical constitution of the beings—

I. *General Morphology*—Four processes: 1) Cellular multiplication; 2) Rejuvenation; 3) Conjugation; 4) Gemmation.

II. *Special Morphology*—Development of the primordial egg—Ovogenic period; theory of encapsulation of germs; epigenesis—Period of fertilization—Embryological period.

Page

III. *Origin and cause of morphology*—Morphology derives from atavism, from the previous state—Distinction between morphological synthesis and chemical synthesis—Final causes; they are merged with primary causes, and have no separate existence 209

NINTH LECTURE

Résumé of the Course

I. *Concept of life*—Life is neither a principle nor a resultant; it is the consequence of a conflict between the organism and the external world—Demonstration of this proposition by various detailed expositions.

II. *Concept of living organisms*—Life is independent of a definite organic form—Law of formation of organisms—The organism is formed in view of the lives of the elements—Autonomy of the lives of the elements and their subordination to the whole—Laws of differentiation and of division of labor—Law of organic perfectionment—Morphological unity of the organism—Various demonstrations—Redintegration, cicatrization, etc.—Diverse forms of vital manifestations—Vital phenomena—Functions—Properties.

III. *Concept of the science of physiology*—General and descriptive physiology—Comparative physiology—The problem of physiology: to know the laws of the phenomena of life and act upon the manifestation of these phenomena—Physiology is an active science—Its principle is determinism, like that of all experimental sciences 249

Explanation of the Plate 274

Appendix 277

LECTURES ON THE PHENOMENA OF LIFE COMMON TO ANIMALS AND PLANTS

OPENING LECTURE*

SUMMARY: Inauguration of general physiology at the Muséum—Reasons for the transfer of my Chair from the Sorbonne to the Jardin des Plantes—Today, Physiology is becoming an independent science, separate from Anatomy—It is an experimental science—Definition of the domain of general physiology—Beginnings in France—Development of Physiology in neighboring countries—Laboratory installations—This is not all; it is necessary above all to have a good method and a healthy experimental critique.

As I begin the course in general physiology at the Muséum d'Histoire Naturelle, I believe it is necessary to indicate the circumstances that have brought me to it. The introduction of general physiology into the famous institution that shelters the natural sciences, and the creation of a laboratory attached to the Chair, mark a notable progress in the teaching of experimental physiology. This completely modern science, born in France through the fruitful impetus of Lavoisier, Bichat, Magendie, etc., has remained up to the present, it must be said, almost without encouragement, while it has on the contrary received a great deal in the neighboring countries. The endowment of physiology at home has been out of proportion to its needs, and I am happy to state that the arrangements by virtue of which I have been called to the Muséum d'Histoire Naturelle are the first steps in the satisfaction of requirements that have become obvious.

It is the consideration of these higher interests alone that determined my decision to transfer here the instruction that I had carried out at the Faculté des Sciences since the year 1854, when

Summer semester, 1870. See *Revue Scientif.*, n° 17, 1871.

the *Chair of General Physiology,* of which I was the first encumbent, was created.

In 1867 Mr. Duruy, minister of public instruction, asked me to describe in a report the progress of general physiology in France, and to indicate the improvements that would contribute to its advancement. Although I was ill at that time, I accepted the task. I did my best to compare the development of our science in France and abroad, and I came to this conclusion, French physiology was poorly provided for, but not inadequate; it was only that support for work was lacking, physiological genius had never been wanting. A conclusion of the same kind could, moreover, be applied generally to the greater part of the physical and natural sciences, and numerous and excellent reports published by my colleagues had made this situation fully evident.*

Rightly concerned and desiring to remedy this state of affairs, Mr. Duruy founded the École Pratique des Hautes Études; at the same time the minister proposed for me the directorship of a public laboratory of physiology in this institution. The state of my health and other considerations at first made me decline this honor; but the minister insisted in the name of science, and I believed that it was my duty to accede to this honorable insistence. It was agreed that my chair at the Sorbonne would be transferred to the Jardin des Plantes, in place of the Chair of Comparative Physiology, which would without doubt be reestablished later. The problem of comparative physiology being the study of the mechanism of life in the various animals, this science has an outstanding place in an establishment which offers, in this regard, resources as complete as the Muséum d'Histoire Naturelle of Paris.

I do not therefore need to continue here the traditions of a predecessor; in reality I inaugurate the teaching of general physiology that I have professed for sixteen years at the Sorbonne.

At the Muséum we shall have a special laboratory and an installation that we lacked at the Faculté des Sciences. I propose today to show you in a rapid way that these new means of study have been rendered indispensable by the very evolution of physi-

*See *La Collection des Rapports.* Paris, Hachette, 1867.

ological science, which demands a growing experimental perfection to attain its goals and to resolve the problems that are incumbent upon it.

Physiology is the science of life; it describes and explains the phenomena characteristic of living beings.

Defined thus, physiology has a problem which is peculiar to itself, and which belongs only to it. Its point of view, its goals, its methods make it an autonomous and independent science; this is why it must have the proper resources for its cultivation and development.

It is necessary to have a sound understanding of the general movement taking place before our eyes, which tends toward the emancipation of physiological science and toward its definitive constitution. It must be said that this evolution appears to have remained unseen by many people who maintain that physiology is a dependency or a part of zoology and of phytology, under the pretext that zoology embraces the whole of the history of animals and that phytology includes the whole of the history of plants. Nevertheless, one does not see the mineralogists contest the independence of physics or chemistry; and indeed they would have as much reason for proclaiming the existence of a single science of inanimate objects, as the naturalists have of proclaiming the existence of a single science of animals, which would be zoology, or a single science of plants which would be botany. The sciences, at first indistinguishable, were not constituted solely according to the more or less natural limits of the objects studied, but also according to the ideas that regulated their study. They are separated not only by their object, but also by their point of view or by their problems.

In the beginning physiology was included in anatomy, and it had no other laboratory than the dissecting amphitheater. After the organs were described, inferences concerning their use were derived from their description and their relationships. Little by little, the physiological problem was disengaged from the anatomical problem, and the two sciences had finally to be separated because each of these sciences pursued a special goal.

Although the development of physiology, which has ended

today in its independence, was progressive, and so to speak unnoticed, we are able nevertheless to distinguish two principal periods in its evolution. The first begins in antiquity with Galen and ends with Haller. The second begins with Haller, Lavoisier, and Bichat, and continues today.

In the first period physiology does not exist as an independent science; it is associated with anatomy, of which it seems to be a simple corollary. Functions and use are judged by the topography of the organs, by their form, by their connections, and by their relations, and when the anatomist calls vivisection to his aid, it is not at all to explain functions, but rather to localize them. It is established that a gland secretes, that a muscle contracts; the problem seems to be resolved and no explanation is demanded; there is a word for everything: it is the result of *life*. Parts are removed, they are ligated, they are cut out, and from the changes that take place in the phenomena, the role attributed to these parts is decided. From Galen until our times, this method has been practiced to determine the function of organs. Cuvier preferred to this method the deductions of comparative anatomy.*

Before the creation of general anatomy, the microscopic elements of the organs and the tissues were unknown, and there could be no question of the intervention of the physicochemical properties of these elements as the agents of the vital manifestations. A mysterious vital force sufficed to explain everything; the name alone changed; according to the times it was called ψυχή, *anima, archeus, vital principle,* etc. Although attempts were made in various ways to explain the vital phenomena by physicochemical actions, nevertheless the anatomical method continued to dominate. Haller, who closes the period about which we are speaking, and who opens the new era, has well summarized the anatomical discoveries, the ideas, and the acquisitions of his predecessors in his immortal *Traité de Physiologie.*

The second period opens, as we have said, at the end of the last century. At this time, three great men, Lavoisier, Laplace, and Bichat, lifted the science of life out of the anatomical rut where it threatened to languish and gave it a decisive and lasting

*See Lettre à Mertrud; *Leçons D'anatomie Comparée, year VIII.*

direction. Thanks to their work, the early confusion of anatomy and physiology tended to disappear, and it began to be understood that the descriptive knowledge of the animal organization was not sufficient to explain the phenomena that took place therein. Descriptive anatomy is to physiology what geography is to history, and just as it is not enough to know the topography of a country to understand its history, so also it is not enough to know the anatomy of organs to understand their functions. An old surgeon, Méry, familiarly compared the anatomists to those messengers one sees in large cities who know the names of the streets and the numbers of the houses, but do not know what takes place within. In fact, in the tissues and in the organs, vital phenomena of a physicochemical nature take place which anatomy cannot account for.

The discovery of respiratory combustion by Lavoisier has been, it can be said, more fruitful for physiology than most anatomical discoveries. Lavoisier and Laplace established this fundamental truth that the material manifestations of living beings are included among the ordinary laws of general physics and chemistry. These are chemical actions (combustion and fermentation) which preside over nutrition, and produce the heat within the organism that maintains the fixed temperature of higher animals. Anatomy can teach us nothing about this subject; it can at most localize these manifestations but not explain them.

On the other hand, by founding general anatomy and relating the phenomena of living bodies to the elementary properties of tissues, like effects to their causes, Bichat established the true and solid base on which general physiology is founded; not that the vital properties were considered by Bichat as special physicochemical properties which leave no further place for the mysterious agents, animism, and vitalism, since his work consisted solely in a decentralization of the vital principle. He localized the phenomena of life in the tissues, but he made no progress along the road of their true explanation. Bichat with Stahl and the vitalists still believed in the antagonism of vital phenomena and physicochemical phenomena; the studies and the discoveries of Lavoisier contained, as we shall see, the refutation of these erroneous ideas.

In résumé, physiology has presented two successive phases; at first anatomical, it has become physicochemical with Lavoisier and Laplace. Life was at first centralized, its manifestations considered as the activities of a single vital principle; Bichat decentralized it, dispersing it throughout the anatomical tissues.

Nevertheless, these ideas of vital decentralization penetrated into science only with difficulty.

In this century there were still experimenters who sought the seat of the vital force, the place where it resided and from which it extended its domination over the entire organism. Legallois experiments to apprehend the seat of life, and he places it in the nervous centers, within the medulla oblongata. Flourens restricts the vital principle within a more circumscribed space, which he calls the vital node. According to the ideas of Bichat, on the contrary, life is everywhere, yet no place in particular. Life is neither a being, nor a principle, nor a force residing in a part of the body, but simply the general consensus of all the properties of the tissues.

After Lavoisier and Bichat, physiology was thus in some way constituted, putting forth two powerful roots, one into the ground of physics and chemistry, the other into the ground of anatomy. But these two roots developed separately and in isolation through the efforts of the chemists who succeeded Lavoisier and the anatomists who followed Bichat. I think that from now on they ought to unite their sap to feed a single trunk and nourish a single science, the new physiology.

Up to this point, the newborn physiology lacked a home of its own, and asked for hospitality from both the chemists and the anatomists.

Nevertheless, Magendie, pushed in the direction of physiology by the advice of Laplace, continued the healthy traditions which he had derived from his contacts with this famous scientist. He introduced experimentation into physiological investigation; he expected from it alone, for the science that he cultivated, the benefits that the physical and chemical sciences themselves had derived from the same method. There had indeed been physiological experimenters in France: Petit (of Namur), Housset, Legallois, and Bichat himself. But by his perseverance, despite all opposition and the greatest of difficulties, Magendie succeeded in bringing about the triumph of the method that he advocated. It is to him that the honor is owed of having exerted a decisive influence on the progress of physiology and of having definitively made it dependent upon experimentation.

It is not without value to recall that while this movement of ideas was taking place in France, the neighboring nations, who have known so well how to profit from it, brought no support to this progress. Germany slept or dreamed in the clouds of natural philosophy; it discussed the legitimacy of experimental knowledge and lost itself in the abstractions of the *a priori* method England did not follow us except at a distance.

It is therefore from our country that the impetus came; and if the movement of renovation did not develop here at all, while it spread to Germany where it brought forth all its fruits, we can at least claim the honorable role of having been its initiators.

Magendie himself had only the most restricted resources at his disposition. He gave private courses in experimental physiology, based on vivisection. It was only after 1830, after having been appointed professor of medicine at the Collège de France, that he established the very inadequate laboratory which still exists there today, and which was the only official laboratory France had at that time. When it began, this experimental teaching of Magendie was, moreover, unique in Europe; many students attended, among them many foreigners who there became involved with the ideas and methods of experimental physiology.

Through his relations with Laplace, Magendie, who was an anatomist, found himself committed to the course of modern physiology which tends to analyze the phenomena of life in terms of physical and chemical explanations; thus, Magendie is the first physiologist to write a book on the *physical phenomena of life.*

Since Magendie was my master, I have the right to take pride in my scientific ancestry, and I have the obligation of trying, to the limit of my ability, to follow the work which is associated with the names of the illustrious men that I have cited.

Having succeeded Magendie at the Collège de France, like him I fought against the lack of resources; I maintained the laboratory at the Collège de France in the face of difficulties, when they wished to close it on the erroneous pretext that medicine was not an experimental science. Despite the scanty means I had at my disposal, I took in numbers of students who today are professors of physiology or of medicine in various universities in Europe or in the New World. At that time the laboratory of the Collège de France was the only one in existence. Since then, splendid installations have been devoted to physiology and experimental medicine in Germany, Russia, Italy, Hungary and Holland; the laboratory at the Collège de France, which was in our country the cradle of physiology and experimental medicine, has not yet been the object of the improvements to which its past has given it so much right.

Finally, physiology is a science that today has become distinct and autonomous, and for it to take shape and develop, it must have a home of its own, separate from those of the anatomists and the chemists. Since it has a particular and well defined problem, it must possess special resources to pursue its study.

The advancement of all sciences takes place along two distinct lines: first from the impetus of discoveries and new ideas, and second through the power of the resources for work and scientific development, in a word, by the cultivation that makes the seeds created by inventive genius bear the fruits hidden within them. In the beginning, as we have already said, when physiology was only an appendage of anatomy, the dissecting amphitheatre was the laboratory for both. After Lavoisier and Laplace, physics and chemistry entered into the study of the phenomena of life, and experimenters had to employ the instruments and apparatus of physics and chemistry. As the science progresses, the necessity is felt more and more for special installations where the requisite tools are assembled for experiments in physics, chemistry, and on living animals, by which physiology penetrates into the depths of the organism. The method that ought to direct physiology is the same as that in the physical sciences; it is the method that belongs to all experimental sciences; it is still today what it was at the time of Galileo. Finally, most questions in science are resolved by the invention of appropriate tools; the person who discovers a new procedure or a new instrument often does more for experimental physiology than the most profound philosopher or the most powerful generalizing mind. Thus the attempt has been made to extend the power of the instruments of research further and further. To attain this goal, institutes of physiology abroad have had to make sacrifices.

The value of special physiological laboratories need no longer be proven by arguments; it is established by the facts. This is recognized by the entire scientific world, and it will suffice for me to enumerate here the establishments of this kind set up abroad, where the chairs of anatomy and physiology, everywhere combined twenty years ago, are now everywhere separate.

Johannes Mueller formerly taught anatomy and physiology at the University of Berlin; the dual regime was introduced a long time ago, and anatomy is now confided to Reichert, and physiology to du Bois-Reymond.

In Würzburg, Kölliker in the beginning taught microscopic anatomy and physiology; he has retained anatomy, and physiology was given to Ad. Fick.

In Heidelberg the teaching of the anatomist Arnold was also divided; Arnold remained an anatomist, and physiology was entrusted to the illustrious Helmholtz.

In the little University of Halle the teaching of Volkmann still remains undivided; this is an exception that will not take long to disappear.*

In Copenhagen physiology is represented by Panum, well known for his studies on blood, by his studies on teratological embryogenesis, and by many other important works.

Scotland has followed the example of Denmark; in Edinburg Bennett will retain only his Chair of Anatomy during the next semester, and physiology will constitute a separate subject.

On all sides they yield to the evidence, and this transformation has become a considerable element of progress. In my report of 1867 I insisted on the value of this separation and pointed out that since France was the starting point of this scientific movement, it was to her honor and interest not to be left behind.

In addition, Mr. Wurtz, Dean of the Faculté de Médecine, was sent to Germany to visit the laboratories there. As a chemist he gave much consideration to chemistry but nevertheless de-

---

*Today this separation has taken place.

voted serious attention to the Institutes of Physiology. He visited in turn the Institute at Heidelberg directed by Helmholtz, that at Berlin entrusted to du Bois-Reymond, that at Goettingen where Rudolph Wagner formerly worked, and which today has the physiologist Meissner as its head. He could not overlook establishments of the same kind at Leipzig and Vienna, the one placed under the eminent direction of Ludwig, and the other under that of Brücke. The physiological institute at Munich under the direction of Pettenkofer and Voit attracted his special attention; he was able to see in this establishment a magnificent apparatus designed to study the products of respiration, a vast and fine installation in which one could, hour by hour, day by day, measure combustion and make an exact balance of the chemical phenomena of life.

Germany has not proceeded in this direction alone; St. Petersburg has fine physiological institutes. In Holland the cities of Utrecht and Amsterdam have justly entrusted the teaching of physiology to Donders and to Kühne.* In Florence and at Turin the same honor has been reserved for Moritz Schiff,† for Moleschott, etc.

I show you the plans for one of these laboratories; it is that at Leipzig directed by Ludwig, which is depicted here in the excellent report by Wurtz. I wish that you could see by this example the wealth of these scientific installations of which we have hardly an idea in France. In the basement are the cellars, the rooms for research at constant temperature, apparatus for distillation, a steam engine which provides motive power everywhere, the workshop for a mechanic attached to the laboratory, a storeroom for chemical products, and a hospital for the dogs. On the first floor are situated the laboratories for animal experimentation and for physics and biological chemistry, the room where mercury is used, rooms for microscopes, for histological studies, for the spectroscope, etc.* The library, the lecture hall, and the

*Today, Kühne is in Heidelberg in the chair occupied before him by Helmholtz.

†Schiff is now in Geneva.

*It is very important for good experimental economy to have separate rooms for experiments that require special instrumentation. In this way one avoids all the loss of time required for a new set-up and for reassembling materials that are

→

lodgings for the professor are part of the same building; if we add to this a stable, an aviary, and numerous aquaria, we shall have enumerated the essential components of this magnificent establishment dedicated to science.

When his laboratory was opened, Professor Ludwig delivered an address in which he emphasized the value of the practical exercises in experimentation, for which he is richly endowed; du Bois-Reymond, Kühne, and Czermak expressed themselves in the same vein, and I, myself, am only the echo of the physiological movement that is taking place everywhere.†

The laboratory of the physiologist is necessarily complex, by reason of the complexity of the phenomena that are studied therein. It is naturally arranged for three different kinds of study: (1) study by *vivisection,* (2) *physicochemical* studies, and (3) *anatomicohistological* studies. If, for example, the question arises of studies on digestion, one must first produce a gastric or pancreatic fistula, etc., by vivisection, then proceed to chemical analysis of the secretions, and finally to get a clear idea of the intimate structure of the glands that secrete these digestive juices. In a word, it is necessary to penetrate into the depths of the organism by a more and more intimate analysis, and arrive at an understanding of the elementary organic conditions which will explain to us the real mechanism of the vital phenomena.

To carry the physiological and physicochemical investigation

†Since the time (1870-71) when this lecture was prepared and published, many new physiological installations have been established. In Hungary laboratories have just been built that exceed, they say, everything that has been done before. In Geneva there are also splendid institutes. France alone, which had nevertheless taken the initiative in this science which will be the honor of the 19th century, remains retarded; although improvements have been made, they are still quite insufficient. We do not wish to say that French physiology has on that account declined; it always holds its honorable place in the scientific world. Although it is useful to have great and handosme laboratories, this is not enough to make great discoveries; it is necessary to establish a healthy physiological critique, to follow a good method, and to have good principles. In a word, there must be a good instrument and a skillful worker.

---

sometimes hard to get back together. This arrangement, which is fundamentally nothing but good use of time, might, moreover, be extended to all scientific endeavors.

of the living body down to its finest subdivisions, and into its most secret recesses, such is the problem we have to resolve. You see the experimental difficulties that arise before us, and you understand the importance of operative procedures, and the utility of instrumentation; in short, the need for a laboratory for this kind of research.

The only way to get at the truth in the science of physiology is by way of the experiment; if we can advance it only slowly thereby, we ought not to be discouraged despite the obstacles and the difficulties, always recalling the words of Bacon, "A cripple walks faster on a good road than a skillful runner on a bad one."

After having insisted on the necessity of being properly installed to pursue the experimental method in physiology, we should conclude by a general remark.

Thanks to the new means of study and to the progress of experimentation itself, studies have been multiplied infinitely within the past few years; today it is less important to increase the number of physiological experiments than to reduce them to a smaller number of decisive demonstrations.

The science of living beings has found its way; it is definitively experimental, which is a considerable step forward; it is now a question of perfecting the method, of giving it all the fecundity that is within it, and making it bear its full fruit by regulating its applications. This cannot take place except by submitting experimentation to a rigorous discipline.

This requirement will be understood by all who follow the development of physiology in its daily progress. The field is already beset by the multitude of studies that often demonstrate more zeal than genuine understanding of the experimental method. It is urgent that critical judgement be applied to these incoherent materials to bring them back to the state of exactitude that physiological experiments demand.

Studies of the phenomena of life are subject to great difficulties. The physiologist must appreciate all the conditions of an experiment in order to know whether he has provided for all of them and discerned those that have varied from one experiment to the other.

When experimental conditions are identical, in physiology as well as in physics or in chemistry, the result is unequivocal; if the result is different, it is because some condition has changed. Thus there is by no means any lack of exactitude in the phenomena of life compared to the phenomena of inanimate objects; it is the experimental conditions which are more numerous, more delicate, and more difficult to recognize and maintain. It is not that life or the influence of some capricious agent intervenes; it is solely the complexity of the phenomena that makes them more difficult to grasp and determine.

The principles of experimentation applied to living creatures can be revealed only by long and steadfast endeavor. To overcome the difficulties of experimental critique and succeed in recognizing all the conditions of a physiological experiment, one must have felt one's way for a long time, have been mistaken thousands and thousands of times, in short, have grown old in the practice of experimentation.

FIRST LECTURE

SUMMARY: I. Definitions in the sciences; Pascal. Definitions of Life: Aristotle, Kant, Lordat, Ehrard, Richerand, Tréviranus, Herbert Spencer, Bichat. Life and Death are two states that are understood only by their opposition—The definition in the Encyclopedia—Life can be characterized but not defined—General characteristics of life: organization, reproduction, nutrition, development, aging, disease, death—Attempts at definition derived from these characteristics—Dugès, Béclard, Dezeimeris, Lamarck, Rostan, de Blainville, Cuvier, Flourens, Tiedemann—The essential characteristic of life is organic creation.

II. Hypothesis of Life: Spiritual and materialistic hypothesis; Pythagorus, Plato, Aristotle, Hippocrates, Paracelsus, van Helmont, Stahl; Democritus, Epicurus, Descartes, Leibnitz—The School of Montpellier—Bichat, etc.—We do not accept either materialistic or spiritualistic hypotheses in physiology because they are inadequate and alien to experimental science—Observation and experiment teach us that the manifestations of life are neither the product of matter nor of an independent force; that they result from the natural interaction of preestablished organic conditions, and precise physicochemical conditions—We can apprehend and understand only the material conditions of this interaction, that is to say the determinism of the manifestations of life—Physiological determinism encompasses the problem of science of life; it will permit us to master the phenomena of life, as we control the phenomena of inanimate objects whose conditions are known to us.

III. Determinism in Physiology—It is absolute in Physiology as in all experimental sciences—It was wrongly attempted to exclude determinism from the science of life—Distinction between philosophical and physiological determinism—Answer to the philosophical objections; physiological determinism is an indispensable

CONDITON FOR MORAL FREEDOM INSTEAD OF BEING ITS NEGATION—NECESSARY SEPARATION OF PHYSIOLOGICAL QUESTIONS FROM PHILOSOPHICAL OR THEOLOGICAL QUESTIONS—THERE IS NO POSSIBLE CONCILIATION BETWEEN THESE SEPARATE PROBLEMS; THEY ARE DERIVED FROM DIFFERENT INTELLECTUAL NEEDS AND ARE RESOLVED BY OPPOSING METHODS—NEITHER THE ONE NOR THE OTHER CAN PROFIT BY A RAPPROACHEMENT.

I. Since physiology is the science of the phenomena of life, it has been thought that this definition implied another, that of life itself. This is why a great number of definitions of life are to be found in the works of the physiologists of all times.

Should we imitate them, and do we believe it is necessary to begin our studies by an enterprise of this kind? Yes, we shall begin like them, but with the quite different purpose of proving that the attempt is chimerical, alien, and useless to the science.

Speaking of the perfect scientific method, in his reflections on geometry, Pascal stated that no term should be employed unless its meaning had been clearly explained beforehand; it should consist in defining everything and proving everything.

He noted immediately however that this is impossible. True definitions are in reality, he said, only *definitions of words,* that is to say the imposition of a name upon objects by the mind, for the purpose of curtailing discussion.

There is no definition for objects that the mind has not created, and which it does not encompass completely; in short, there are no definitions of *natural objects.* When Plato defines *man* as a "two-legged animal without feathers," far from giving us a clearer understanding than before, according to Pascal he gives us a useless and even ridiculous idea, since, he adds, "a man does not lose his humanity by losing his legs, nor does a capon acquire it by losing its feathers."

Geometry can define the objects of its study, because they are a pure creation of the mind: definition is thus a convention that the mind is free to establish. When an *even number* is defined as a "number divisible by two," a geometrical definition is given, according to Pascal, because a name is employed, divested of all other meanings, if it has any, to give it that of the thing designated.

The procedure is the same in philosophy, because it deals mostly with intellectual concepts; and even there, there are fundamental terms that cannot be defined.

The same thing also happens in geometry, where the basic notions of *space,* of *time,* of *movement,* and the like, are not defined. They are employed in conversation without confusion because people have a sufficient knowledge and a clear enough idea of them, so that they are not mistaken about the thing designated, however obscure the idea of this thing can be when considered in its essence. This comes about, Pascal also says, because nature has given all men the same primitive ideas about primitive things. This is what the celebrated mathematician Poinsot recalls humorously, "If someone asks me to define *time,* I would reply to him: 'Do you know what you are talking about?' If he tells me, 'Yes'—Very well, let us talk about it. If he says, 'No'—Very well, let us talk about something else."

When one wants to define these basic notions, one can never clarify them by anything more simple; one is always obliged to include in the definition the very word to be defined. "Time is a *succession* . . . ." said Laplace. But what is a succession, if one does not already have the idea of time? Do not these definitions recall the one Pascal ridiculed: "Light is a luminous motion of luminous bodies."

Nothing can be defined in the natural sciences; all attempts at definition express nothing more than a simple hypothesis. Objects are known only in succession, from different and diverse points of view. At the beginning of these sciences, no one possesses a detailed and complete understanding of them such as a definition presupposes; this comes at the end, and as an ideal and inaccessible end-point of the study.

The method that consists in defining and deducing everything from a definition can serve the sciences of the mind but it is contrary to the very spirit of the experimental sciences.

This is why there is no need to define life in physiology. When one speaks of life, one is understood on the subject without difficulty, and this is enough to justify the use of the term in an unequivocal way.

It is enough to agree on the word *life* to employ it; but above all it is necessary for us to know that it is illusory and chimerical and contrary to the very spirit of science to seek an absolute definition of it. We ought to concern ourselves only with establishing its characteristics and arranging them in their natural order of rank.

It is important today to clearly liberate general physiology from the illusions that have agitated it for such a long time. It is an experimental science and does not have to give *a priori* definitions.

If after these preliminaries we recall nevertheless the principal attempts at a definition of life made at various times, it will be to show their inadequacy or errors. This study will moreover have another interest for us; it will help us to find, through analysis of all these intellectual efforts, the best concept that we could formulate today about the phenomena of life.

Aristotle says that, "Life is nutrition, growth, and decline, having as its cause a principle that has its complete expression within itself, e.g. entelechy." Now it is this principle that should be grasped and understood.

Burdach recalls that for the *philosophy of the absolute,* "life is the soul of the world, the equation of the universe." He says further that "in life matter is only an accident, while activity is its substance." We shall not dwell on considerations so transcendental that they hold nothing tangible for the physiologist.

Kant has defined life as "an internal principle of action." In his appendix on teleology, or the science of final causes, he says: *"The organism is a whole resulting from a reasoning intelligence residing within its interior."*

This definition, which recalls that of Hippocrates, was accepted in a more or less modified form by a great number of physiologists. But as we shall see later the reason that led to its adoption is fundamentally specious or apparent only. The principle of action of living beings is not within them; it cannot be separated or isolated from external atmospheric or cosmic conditions, and there is no single phenomenon that can be attributed exclusively to it. The spontaneity of the vital manifestations is only an illusion soon disproved by study of the facts. There are always external agents, outside stimulants, that provoke the manifestations of the properties of a material that is always inert by itself. In the higher creatures these stimulants in reality reside within what we call an *internal environment,* but this environment, although deeply situated, is still external to the elementary organized particle which is the only part that is really living.

Lordat accepts a vital principle when he says that "life is the temporary alliance of the inner sense and of the material aggregate cemented by a ένορμὸν or a cause of motion which is unknown to us."

Tréviranus, like Kant, had in mind the apparent independence of vital manifestations from external circumstances. "Life is," for him, "the constant uniformity of phenomena in the face of the diversity of external influences."

Müller seems to accept a sort of vital principle. According to him, there are two things in the seed: the material of the seed, plus the vital principle.

Ehrard considers life as a motor principle; "the faculty of motion destined for the service of that which is moved."

Richerand recognizes implicitly the existence of a vital principle as the cause of a limited sequence of phenomena in living beings: "Life," he says, "is a collection of phenomena which follow each other over a limited time in organized bodies."

Herbert Spencer has more recently proposed a definition of life that I have cited in an article in the *Revue des deux Mondes* (Vol. IX, 1875) in a way that brought protests from the English philosopher. On page 709 of the French translation of his *Principles of Psychology,* we read this phrase:

"Thus, in its final form we shall put forward as our definition of life, *the definite combination of heterogeneous changes, at once simultaneous and successive.*"

This definition, which I had quoted in full, should be completed, it appears, by the addition of these words: *corresponding to external coexistences and sequences.*

According to Mr. Cazelles, the translator of Herbert Spencer, who expressed this criticism (*Revue Scientifique,* No. 33, February, 1876) , the intent of the philosopher would be distorted without the addition of the second part of the phrase. The definition was thus made at different times, by successive degrees, and this procedure, which is not customary, is quite capable of misleading the reader.

In résumé, the translator adds, the essential feature by which Herbert Spencer chooses to define life is *the continuous accommodation of internal relations to external relations.*

Bichat proposes for us a more physiological and more comprehensible idea. His definition of life has had a resounding fame: *"Life is the ensemble of the functions that resist death."*

Bichat's definition includes two terms that are in opposition to each other: *life, death.* It is impossible, in fact, to separate these two ideas; what is living will die; what is dead has lived.

But Bichat wanted to be more clear: he delved into the problem more deeply and encountered the error within it. In a way he had made two entities of life and death, two principles continually present and in conflict within the organism. However much he repudiated the *vital principle* as a unique principle, he gives us its equivalent in his vital properties. These subordinate vital principles, these vital properties, are the agents of life; on the contrary, the physical properties that are in conflict with them are so to speak the agents of death.

All Bichat's contemporaries shared his point of view and paraphrased his formula. Pelletan, a surgeon at the Paris school, teaches that life is the resistance opposed by organized matter to the causes that ceaselessly tend to destroy it. In an often quoted passage Cuvier himself develops this thought, that life is a force that resists the laws controlling inanimate matter; death is the lack of the principle of resistance, and the cadaver is nothing but the living body fallen back under the dominion of physical forces.

Thus according to Bichat, physical properties are not only alien to the manifestations of life, and are to be neglected in its study, but more than that, they are antagonistic to them.

These ideas of antagonism between the general external forces and the internal or vital forces had already been expressed by Stahl in an obscure and almost barbarous language; expressed by Bichat with a luminous clarity, they seduced and captivated every mind.

Science, it must be said, has condemned this definition, according to which there would be two kinds of properties in living beings: physical properties and vital properties, constantly in conflict and the one tending to predominate over the other. In fact, it would result logically from this antagonism that the more the vital properties predominate in an organism, the more the physicochemical properties in it would have to be attenuated, and reciprocally, that the vital properties would have to exhibit more weakness as the physical properties acquired greater strength. Now, it is the opposite that is true: discoveries in physics and biological chemistry have established, in the place of this antagonism, an intimate accord and a perfect harmony between vital activities and the intensity of the physicochemical phenomena.

In sum, the concept of Bichat involves two ideas; the first establishing a necessary relation between life and death, and the second postulating an opposition between vital phenomena and physicochemical phenomena.

The latter part is erroneous.

As to the first, it had already been expressed more simply in its definition in the *Encyclopédie* in a form that almost makes a naïveté of it: "Life is the opposite of death."

This is in fact because we cannot distinguish life except by death, and the reverse. By comparing the living body with the same body as a cadaver, we detect the disappearance of something that we call life.

The citations we have given above show us an apparently great variety in the definitions of life; nevertheless, they all have a common basis that is precisely what constitutes their fault. Nearly all authors have admitted, implicitly, or explicitly, that the manifestations of life have as their cause a *principle* that gives rise to them and directs them. Now to admit that life derives from a vital principle, is to define life by life; this is to introduce what is to be defined into the definition.

It is true that other physiologists have agreed, without giving any better definitions of it, that life instead of being an immaterial directing principle is only a *resultant* of the activity of the organized matter.

Thus, for *Béclard,* life is organization in action.

For *Dugès,* life is the special activity of organized beings.

For *Dezseimeris,* life is the mode of existence in organized bodies.

For *Lamarck,* life is a state of things that permits organic movement under the influence of excitants.

This *state of things* is evidently organization, together with the condition of sensibility.

Rostan, who had located the characteristic of life in organization, and had formulated *organicism,* expresses himself in the following terms:

> The Creator does not communicate a force which he adds to the organized being, since he has placed in this being along with its organization the molecular disposition appropriate for its development. It is the clockmaker who has constructed the clock, and in assembling it has given it the ability to pass through successive phases, to mark the hours, the minutes, the seconds, the phases of the moon, the months of the year, all this over a greater or shorter length of time, but this ability is nothing other *than that which results from its structure;* it is not a separate property, a superadded quality, it is the assembled machine.

Life is the *assembled machine:* the properties derive from the structure of the organs. This is *organicism.* Nevertheless this concept has something vague about it: structure is not a physicochemical property, nor a force that can be the cause of anything by itself, for it in its turn would presuppose a cause.

In summary, all *a priori* views of life, whether considered as a *principle* or as a *result,* have afforded nothing but inadequate definitions, and this has to be so, because the phenomena of life can only be known *a posteriori,* like all the phenomena of nature.

The *a priori* method is thus stamped with sterility, and it would be a waste of time to continue to seek the progress of physiological science in that direction.

Renouncing therefore the definition of the undefinable, we

shall simply try to characterize living beings in relation to inanimate objects. This way of conceiving the problem will lead us to formulations which will express the facts, and no longer only ideas or hypotheses.

It is not that we reject hypotheses in science; in any event they are only the scaffolding; science is constituted by facts, but it makes progress and grows with the help of hypotheses.

Let us now examine the general characteristics of living beings. They can be reduced to five, namely:

Organization
Reproduction
Development
Old age, illness, death
Nutrition

A. *Organization* results from a composite of complex substances reacting one upon the other. For us this is the arrangement that gives rise to the immanent properties of living matter, an arrangement that is special and very complex, but which nevertheless obeys the general chemical laws of the association of matter. Vital properties are in reality only the physicochemical properties of organized matter.

B. The faculty of *reproduction* or of *generation*, that is to say the act by which beings are derived one from the other, characterizes them in an almost absolute manner. Every being comes from parents, and at a certain time becomes capable of being a parent in its turn, that is to say to give rise to other beings.

C. *Development* is perhaps the most remarkable trait of living beings, and consequently of life.

The living being appears, grows, declines, and dies. It is continually changing: it is subject to death. It comes from a germ, an egg, or a seed, attains a certain degree of development through successive differentiations, and forms organs, some short-lived and transitory, the others having the same duration as itself; then it destroys itself.

The inanimate mineral object is immutable and incorruptible, as long as external conditions do not change.

This characteristic of a determined development, of a beginning and of an end, of continuous progress in one direction within a fixed term, belongs inherently to living beings.

In truth, astronomers today accept the idea of mobility and continuous evolution of the sidereal world. But in this possible evolution of sidereal bodies, compared to the rapid development of living beings, there is a difference in degree which from the practical point of view is enough to distinguish them. Relative to us, the world and the stars evince insensible changes only: living beings on the contrary exhibit an appreciable development.

Death is also a necessity to which the living individual is inevitably subject, and thereby returns to the mineral world. He is moreover subject to *disease,* and capable of recovery. The medical and naturalist philosophers have been vividly struck by this tendency of the organized individual to reestablish its form, to repair its mutilations, to heal its wounds, and in this way to prove its unity, and its morphological individuality.

According to certain physiologists, this tendency to realize and to repair a sort of individual architectural plan would shape the organized being into a harmonic whole, a sort of small world within the large; this then would be an exclusive characteristic of bodies endowed with life. "Inorganic bodies," says Tiedemann, "exhibit absolutely no phenomena that can be considered as the effect of regeneration or of healing. No crystal reproduces the parts that it has lost, nothing repairs the breaks that have occurred in its continuity, nothing returns by itself to its state of integrity."

This is not correct; crystals, like living bodies, have their form and their particular plan, and when disturbing actions within the ambient milieu displace them from them, they are capable of reestablishing them by a true *healing* or *redintegration* of the crystal. Pasteur has seen "that when a crystal has been broken in any one of its parts, and when it is replaced in its mother liquor, it is seen that at the same time that the crystal increases in all its dimensions by a deposition of crystalline particles, a very active work takes place on the broken or deformed parts; and in several hours it has accomplished, not only the regularity of the general work on all parts of the crystal, but also the reestablishment of regularity in the mutilated part . . ." Thus the physical force that arranges the particles of a crystal according to the laws of a wise geometry has results analogous to those which arrange the organized substance in the form of an animal or a plant. This characteristic is not therefore as absolute as Tiedemann thought it to be; nevertheless, it has at least a degree of intensity and of energy peculiar to living beings. Moreover, as we have said, there is in the crystal none of the development that characterizes the animal or the plant.

D. Finally, *nutrition* has been considered as the distinctive and essential trait of living beings, the most constant and the most universal of its manifestations, and the one which in consequence ought and could (by itself) suffice to characterize life.

Nutrition is the continuous mutation of the particles which constitute the living being. The organic edifice is the site of a perpetual nutritive movement which leaves no part at rest; each one, without cease or respite, takes its food from the medium that surrounds it and into it rejects its wastes and its products. This molecular renovation is imperceptible to the sight, but as we see its beginning and its end, the intake and the output of substances, we conceive of its intermediate phases, and we represent to ourselves a flow of material which passes incessantly through the organism and renovates it in its substance and maintains it in its form.

The universality of such a phenomenon in the plant, in the animal, and in all their parts, and its constancy which suffers no interruption, make it a general sign of life, which several physiologists have employed in its definition.

Thus *de Blainville* has said:

"Life is a double internal movement of composition and decomposition at once general and continuous."

Cuvier expresses himself in the same way:

"The living being," he says, "is a whirlpool constantly turning in the same direction, in which matter is less essential than form."

Flourens has paraphrased this idea of a *vital whirlpool* or a material circulus, by saying:

"Life is a form served by matter."

Finally, *Tiedemann,* accepting both the double movement of

composition and decomposition of living beings, connects it to a vital principle which governs it.

"Living bodies," he says, *"have their principle of action within themselves, which prevents them from ever falling into chemical indifference."* The definition derived from this characteristic merits our attention for a moment.

We have already said that the manifestations of life cannot be considered as being regulated directly by an internal vital principle. The activity of animals and plants is certainly subject to external conditions. This is indeed apparent in plants and in cold-blooded animals which go to sleep in the winter and wake up during the heat of the summer. We shall see later that if man and warm-blooded animals seem to be free in their acts and independent of variations in the cosmic environment, this occurs because there exists in them a complex mechanism that maintains around the living particles, fibers, and cells, a truly invariable environment, which is the blood, always of the same warmth and of constant composition. They are independent of the external medium because thanks to this artifice, the internal environment does not change around their active and living elements. In the living animal, in reality, there are always external agents and foreign extracellular stimulants which provoke the manifestations of the properties of a matter that by itself is always both inactive and inert.

If an internal principle existed, and were independent, why should life become more energetic in the summer than in the winter in certain living beings, more vigorous in the presence of oxygen than in its absence, or more active in the presence of water than after desiccation?

It is not correct to say, on the other hand, that living bodies are incapable of falling into a state of chemical indifference. In truth, whatever may be in ordinary circumstances the torpor into which the plant or cold-blooded animal has lapsed, life has not ceased within it; the organism has not really fallen into absolute inertia, or into a genuine state of chemical indifference. But we can prove that this situation is realized in the being in the state of *latent life*. Here is a seed; it is as inert as a mineral body. In certain circumstances its constitution remains invariable and it will remain thus for months and centuries. Is it alive? No, according to the definition of Tiedemann, since this seed is in a state of complete chemical indifference. Nevertheless, if it is given the external conditions for germination, heat, humidity, and air, it will germinate and develop a new plant. We shall show you that it is the same with resuscitated or reviviscent animals, such as rotifers and eel-worms, which can revive after having lapsed, theoretically for an indefinite time, into the most complete inertia.

What is to be concluded from this if not that the vital phenomena are not at all the manifestations of the activity of a free and independent internal vital principle? One cannot apprehend this interior principle, isolate it or act upon it. On the contrary, one always sees vital acts, always having as their condition, external physicochemical circumstances, perfectly determined and capable either of preventing or permitting their appearance.

In résumé, the vital whirlpool is not the unique manifestation of a *quid intus,* nor the sole effect of external physicochemical conditions. In consequence, life cannot be characterized exclusively by a vitalistic or materialistic concept. The attempts that have been made in this direction throughout the ages are illusory and can lead only to error.

Should we stop at this negation?

No. A negative critique is not a conclusion. We must formulate an idea in our own turn and look for a characteristic whose value, although not absolute, will be capable of illuminating our route without ever misleading us.

The characteristics that we have recalled in the preceding correspond to realities; they are good and worth knowing. For my part I shall mention the concept to which my experience has led me.

I consider that in the living being there are necessarily two orders of phenomena:

1. The phenomena of *vital creation* or *organizing synthesis.*
2. The phenomena of death or *organic destruction.*

It is necessary to explain ourselves in a few words about the meaning that we give to these expressions, organic *creation* and *destruction*.

While from the standpoint of inorganic matter, we accept, with reason, that nothing is lost and that nothing is created, it is not the same from the standpoint of the organism. In a living being everything creates itself morphologically and organizes itself, and everything dies and destroys itself. In the developing egg muscles, bones, and nerves appear and take their place, repeating a previous form from which the egg was derived. The ambient material is assimilated into tissues, either as a nutritive principle, or as an essential element. The organ is created, and is so from the point of view of its structure, its form and the properties that it manifests.

On the other hand, organs destroy themselves, disorganize themselves constantly by their very own action; this disorganization constitutes the second phase of the great act of life.

The first of these two orders of phenomena is alone without direct analogy; it is particular, special to the living being; this developmental synthesis is what is truly vital. On this subject I shall repeat the formula that I have expressed for a long time: *"Life is creation."**

The second, vital destruction, is on the contrary of a physico-chemical nature, most often the result of combustion, of fermentation, or of putrefaction; of an action, in a word, comparable to a great number of chemical actions of decomposition or of cleavage. These are the true phenomena of *death* when they are applied to the organized being.

It is worth noting that in this we are victims of a habitual illusion, and when we wish to designate phenomena of *life*, we refer in reality to phenomena of *death*.

We are not greatly impressed by the phenomena of life. The organizing synthesis remains internal, silent, and hidden in its phenomenal expression, collecting silently the materials that are to be expended. We do not see these phenomena of organization directly, only the histologist, and the embryologist, following the development of the element or of the living being, apprehend

*See *Introduction à l'Étude de la Médecine Expérimentale,* p. 161, 1865.

changes, and phases that reveal this silent work to them; here there is a deposition of material, there, a formation of a membrane or a nucleus, over there, a division or a multiplication, a renovation.

On the contrary, the phenomena of destruction or of vital death are those which capture the eyes and by which we are led to characterize life. The signs are evident and obvious: when movement is produced, when a muscle contracts, when volition and sensibility are manifest, when thinking takes place, when a gland secretes, the substance of the muscle, the nerves, the brain, the glandular tissue disorganizes itself, destroys itself and consumes itself. In this way every manifestation of a phenomenon in the living being is necessarily linked to an organic destruction; it is this that I intended to express when in a paradoxical form, I said elsewhere (*Revue des Deux Mondes,* vol. IX, 1875) : *life is death.*

The existence of all beings, animals or plants, is maintained by these two orders of necessary and inseparable acts: *organization* and *disorganization.* As a practical object our science will tend to fix the conditions and the circumstances of these two orders of phenomena.

This classification of vital manifestations which we have adopted is to our mind the very expression of reality; it is the result of the observation of phenomena. To this advantage of being a factual truth, it joins a no less appreciable one of being useful in understanding phenomena, of being profitable in study, of projecting a bright clarity upon the recognition of the modalities of life. It is this that we shall attempt to demonstrate in the course of our lectures; this will be our program.

We have thus arrived, we believe, at the two general facts most characteristic of living beings; this is not enough, the mind has need to go beyond the facts; it feels itself drawn beyond that, and builds hypotheses from which it demands the explanation of things and the means of penetrating them more deeply.

This is why, along with the observation of phenomena, there have always been hypotheses and views expressed regarding life by philosophers, naturalists, and physicians, from the greatest

tendency that seems to have revived in our day to attempt to inject questions of theology and philosophy into physiology and to pursue their supposed reconciliation, is to my mind a sterile and unhappy tendency, because it mixes sentiment and reason, and confuses what one recognizes and accepts without physical demonstration with what ought not to be admitted except experimentally and after complete demonstration. In reality, one can be neither spiritualist nor materialist except by sentiment; one is a physiologist by scientific demonstration.

Philosophy and theology are free to treat the questions that concern them by the methods which belong to them, and physiology intervenes neither to support them nor to attack them. She too has her freedom of action, her particular problems, and her special methods of resolving them. These are thus separate domains in which everything ought to remain in its place; this is the only way of avoiding confusion and insuring progress in the physical, intellectual, political, and moral spheres.

Here we shall be the physiologist only, and as such we cannot place ourselves either in the camp of the vitalists or in that of the materialists. We shall dissociate ourselves from the vitalists because the *vital force*, or whatever name we give it, can do nothing by itself, can only act by drawing upon the action of the general forces of nature, and is incapable of manifesting itself without them.

We shall dissociate ourselves equally from the materialists, for although the vital manifestations remain directly under the influence of physicochemical conditions, these conditions cannot arrange or harmonize phenomena in the order and the succession that they assume in a special way in living beings.

In the face of the phenomena of life we shall remain as men of experimental science, observers of facts, without preconceived systematic ideas. We shall seek to determine exactly the conditions for the manifestations of the phenomena of life so that we can master them, like the physicist and the chemist make themselves masters of the phenomena of inorganic nature.*

Such is the problem of modern physiology, and we certainly

---

*See on this subject: *Revue des Deux Mondes: Problème de la Physiologie Générale.* —*My Rapport sur les Progrès de la Physiologie Générale, 1867.*

would not have arrived at its solution by means of either spiritualist or vitalist doctrines, or with the help of materialist doctrines.

There is an irremediable error at the basis of *vitalist doctrines* which consists in considering as a force a misleading personification of the arrangement of things, in giving a real existence and a material and efficacious activity to something immaterial that is in reality only a notion of the mind, a direction that is necessarily inactive.

The idea of a cause presiding over the sequences of vital phenomena is without doubt the first that presents itself to the mind, and it seems to be undeniable when one considers the rigorously fixed development of phenomena, so numerous and so well concerted, by which the animal and the plant maintain their existence and run their course. Seeing an animal leave the egg and acquire in succession the form and the constitution of the being which precedes it and of that which will follow it, watching it at the same time execute an infinite number of apparent or occult acts which together contribute as by a calculated design to its conservation and maintenance, one has the feeling that a cause directs the concert of its parts and guides in their roles the isolated phenomena for which it is the theater.

It is to this cause, considered as a directive force, that the name of physiological soul or *vital force* can be given, and it can be accepted on condition of defining it and not attributing to it anything except what belongs to it. It is because of a false interpretation that the vital principle has, so to say, been personified and made the instrument of all organic activity. It has been considered as the executive agent of all phenomena, the intelligent agent that models the body and manages the inert and obedient matter of the animate being. For the vitalists the sufficient cause for each act of life lay within this force which had no need of the outside aid of physical or chemical forces, or which even contended with them to accomplish its task.

But experimental science very definitely contradicts this view; it is at this point that it enters into the system to show its fundamental falsity. In fact, physiological research teaches us that the vital force or forces can do nothing without the help of physical conditions. There is an intimate accord, a close liaison between physical and chemical phenomena and the phenomena of life. There is a perfect parallelism, a necessary harmonic union. Humidity, heat, air, create the indispensable conditions for the functioning of life. The vital manifestations are increased or attenuated along with the chemical activities of the tissues, and in proportion to this very action. Lowering the temperature brings on a lowering of sensibility and of intelligence, and produces a depression of life. In desiccation certain beings are plunged into a state of apparent death, which ceases, as we shall see, only when water and the physicochemical conditions necessary to vital manifestations are restored. In such cases should it be said that heat increases the vital force, that cold benumbs it, that desiccation destroys it, and that humidity revives it? But in this case, it would no longer be it which would command the material of the organism, but rather it would be the material state of the organism which would govern it. Indeed, the vital force can accomplish nothing without the physicochemical conditions; it remains absolutely inert, and the vital phenomenon appears only when the fixed physicochemical conditions required for its manifestation are brought together.

This is what the vitalists did not understand, nor did *Stahl,* who confused and identified the *vital force* with the intelligent and reasonable soul; neither did Bichat, who substituted the *vital properties* for this single principle, that is to say, a multitude of *vital forces* residing within each tissue. These vital properties, as he calls them, were opposed to the physical properties; the first being changeable and ephemeral, the second constant and permanent, meeting in the animal body as on a battlefield and struggling without rest or respite, until the moment when, victory resting with the physical agents, the living body died.

Thus, whether vitalism be envisaged in its most extreme expression, as Stahl developed it, or in the milder and more scientific form that Bichat gave to it, it is equally inacceptable, because it is in contradiction with experience and with the facts of physiology.

If, as we have just seen, the vitalist doctrines misunderstood the true nature of vital phenomena, the *materialist doctrines,* on the other hand, have been no less in error, although in an opposite direction.

Accepting that vital phenomena are related to physicochemical manifestations, which is true, the question in its essence is not clarified thereby, for it is not a fortuitous encounter of physicochemical phenomena that fashions each individual according to a plan, and following a design fixed and foreseen in advance, and maintains the admirable subordination and harmonious concert of the acts of life.

Within the living body there is an arrangement, a sort of ordinance, that cannot be left in obscurity because it is truly the most outstanding trait of living beings. We readily agree that the idea of this arrangement is poorly expressed by the word *force,* but here the word is of little importance; it suffices that the reality of the fact is indisputable.

The vital phenomena do indeed have their rigorously determined physicochemical conditions, but at the same time they are subordinated and succeed each other sequentially according to a law fixed in advance; they repeat themselves eternally, with order, regularity, and constancy, and harmonize with each other with the result in view of the organization and the growth of the individual, whether animal or plant.

It is like a preestablished design for each being and for each organ, so that, if considered separately, every phenomenon of the economy is tributary to the general forces of nature; when taken it its relationships with the others it reveals a special linkage, it seems directed by some invisible guide along the path that it follows, and is led to the place that it occupies.

The simplest reflection leads us to recognize a characteristic of the first order, a *quid proprium* of the living being, in this preestablished vital ordonnance.

At any rate observation teaches us only this: it shows us an *organic plan,* but not an active *intervention* by a vital principle. The only *vital force* that we could accept would only be a sort of legislative force, but in no way executive.

To summarize our thoughts, we could say metaphorically, the

vital force directs the phenomena that it does not produce; physical agents produce the phenomena that they do not direct.

Since the *vital force* is not an active executive force, and does nothing by itself, and since everything is manifest in life through the intervention of physicochemical conditions, consideration of this entity should not intervene in experimental physiology. When the physiologist wishes to understand and to evoke the phenomena of life, to act upon them and to modify them, it is not to the *vital force,* an imperceptible entity, to which he must address himself, but to the physical and chemical conditions that control and command vital manifestations.

Whatever subject he studies, the physiologist never finds before him anything but mechanical, physical, or chemical agents. When he examines, for instance, the action of anesthetic substances on sensibility and intelligence, he determines that ether or chloroform act materially and in a physical or chemical way upon the nervous substance, and not at all on a vital principle, nor on a vital function, such as *sensibility,* which is imperceptible by itself. Since it is the same for all the phenomena of life, the physicochemical sciences seem to comprehend within their laws the manifestations of the phenomena of living organisms; from this derives the materialist opinion that life is only an expression of the general phenomena of nature. However this may be, what we know is that the vital principle executes nothing by itself, and that it borrows its forces from the external world in the thousands and thousands of manifestations that appear before our eyes.

From the preceding it results that the conditions that are accessible to us to make the phenomena of life appear are all material and physicochemical. There is no action possible, except *on* and *by* matter. The universe shows no exception to this law. All phenomenal manifestations, whether they reside in living bodies or outside of them, have material conditions as a necessary substratum. It is these conditions that we call the *determined conditions* of a phenomena.

We can understand only the material conditions and not the intimate nature of the phenomena of life. For this reason, we are concerned only with matter and not with primary causes or with a vital directive force derived from them. These causes are inaccessible to us; to believe anything else is to commit an error of fact and of doctrine; it is to be a dupe of metaphor and to take a figurative language as real. In fact, one often hears that the physicist acts on electricity or on light, that the physician acts on life, on health, on fever, or on disease; these are ways of speaking. Light, electricity, life, health, disease, fever, are abstract things that an agent of some kind cannot influence, but there are material conditions which cause the appearance of phenomena which are attributed to electricity, to heat, to light, to health, to disease; we can act upon them and thereby modify these different states.

The concept that we formulate of the goal of all experimental sciences and of its methods is thus general; it belongs to physics and to chemistry and is applied to physiology. It comes back to saying, in other words, that a vital phenomenon, like all other phenomena, has a rigorous determinism. The vital force and life belong to a metaphysical world; their expression is an intellectual requirement, and we cannot use it except subjectively. Our mind seizes upon the unity, the interrelationships, the harmony of phenomena and it considers this as the expression of a *force,* but it would be very wrong to believe that this metaphysical force is active. It is moreover the same with what we call *physical forces;* it would be a pure illusion to attempt to elicit anything by them. These are necessary metaphysical concepts, but they do not leave the intellectual domain in which they were born, and cannot act in any way upon the phenomena which gave the mind the occasion of creating them.

In a word, this developmental, directive, and morphological faculty by which life is characterized is useless in experimental physiology because, being outside the physical world, it cannot exercise any retroactive action upon it. It is thus necessary to separate the metaphysical world from the physical world which serves as its foundation but which has nothing to gain from it, and conclude by paraphrasing the words of Leibnitz: "Everything takes place in the living body as though there were no vital force."

III. The scope and the role of physiology are fixed by the above. It is a science of the same order as the physical sciences; it studies the physicochemical determinism corresponding to the vital manifestations; it has the same principles and the same methods.

In no experimental science is anything known except the *physicochemical conditions* of the phenomena; one does not strive for anything else than to determine these conditions. Nowhere does one attain primary causes; *physical forces* are just as obscure as *vital force* and just as removed from direct access by experiment. One does not act on these entities at all but only on the physical and chemical conditions that bring about these phenomena. The purpose of all natural sciences, in a word, is to establish the determinism of phenomena.

The principle of *determinism* thus dominates the study of the phenomena of life like that of all the other phenomena of nature.

I have expressed this opinion for a long time, but when I employed the word *determinism* for the first time* to introduce this fundamental principle into physiological science, I did not think that it would be confused with the philosophical determinism of Leibnitz.

In any event, if the word *determinism* that I employed is not new, the meaning that I applied to it in experimental physiology is new, and this had to be since Leibnitz had applied it only to purely metaphysical objects while I applied it, on the contrary, to physical objects to characterize the method of physiological science.

When Leibnitz said, "the human soul is a spiritual automaton," he formulated *philosophical determinism*. This doctrine maintains that the phenomena of the soul, like all phenomena in the universe, are rigorously determined by the series of antecedent phenomena, inclinations, judgements, thoughts, desires, and the prevalence of the strongest motives, by which the soul is entrained. It is the negation of human liberty, the affirmation of *fatalism*.

Physiological determinism is quite different. It is the expression of a physical fact. It consists in this principle that each vital phenomenon, like every physical phenomenon, is invariably determined by physicochemical conditions which, permitting it or

---

*See *Introduction à l'Étude de la Médecine Expérimentale*, p. 115, 1965.

preventing it from appearing, become the *conditions* or the *immediate or proximate material causes* for it. The ensemble of the conditions determining a phenomenon necessarily entrain the phenomenon. This is what must be substituted for the ancient and obscure spiritualist or materialist notion of *cause*.

This principle is fundamental in all the physical sciences. There it is incontestable; it does not even need to be affirmed. It is otherwise in the sciences of life. When in fact it becomes necessary to extend the principle of determinism to the facts of living nature, the animist and vitalist physicians and the philosophers oppose it.

The vitalists deny determinism because according to them the vital manifestations have as their cause the spontaneous efficacious action, as though it were voluntary and free, of an immaterial principle. The consequences of this error are considerable; the role of man in the presence of vital facts would be that of a simple spectator, not of an actor; the physiological sciences would be only conjectural, and uncertain. Experiment would have no access to them, observation could not predict them. Here it can be seen, is a doctrine of indolence par excellence; it disarms man. It relegates causes to the outside of objects, it transforms metaphors into substantial entities, and it makes physiology a sort of inaccessible metaphysiology.

Thus it can be seen that the vitalistic doctrine leads necessarily to indeterminism.

This is precisely the necessary conclusion to which Bichat was lead, almost despite himself. When he begins to express his views so clearly and so scientifically in the introduction to his *Anatomie Générale,* one believes that he is going to attach himself solidly to those views that have become the basis of modern science, and repudiate the vitalistic ideas that they contain. Bichat in fact puts forth this general idea, luminous and fruitful, that in physiology as in physics, phenomena should be attached causally to properties inherent in living matter. "The relationship of properties as causes with phenomena as effects is," he said, "an axiom that is almost superfluous to repeat today in physics and chemistry; if my book establishes a similar axiom in the physiological sciences it will have fulfilled its purpose."

But now, after this clear beginning, he distinguishes vital properties from physical properties; the first the agents of life, the others as the agents of death; he places them in conflict, opposing them. Its vital properties are at war with the physical properties as was the *soul* of Stahl. This is just as categorical a negation of determinism in physiology.* Here in fact are the scientific heresies into which Bichat finds himself fatally led:

> Physical properties being fixed and constant, the laws of the sciences which treat of them are equally constant and invariable; they can be predicted and calculated with certainty. Vital properties, having *instability* as their essential characteristic, and all vital functions being susceptible to a host of variations, nothing can be predicted or calculated with regard to their phenomena. From which it is necessary to conclude that absolutely different laws preside over the one and the other class of phenomena.

Elsewhere Bichat says *(Recherches Physiologiques sur la Vie et la Mort, p. 84):*

> Physics and chemistry relate to each other because the same laws preside over their phenomena, but an immense gulf separates them from the science of organized bodies because an enormous difference exists between these laws and those of life. To say that physiology is the physics of animals is to give an extremely inexact idea of it; I might just as well say that astronomy is the physiology of stars.

We could multiply these proofs of the indeterminism or the scientific negation into which, despite his genius, Bichat found himself led by the vitalist doctrines which reigned in these days and from which he could not free himself; but time has already begun to separate error from truth, and since men are great only by the services they render, Bichat will nevertheless live in posterity by the truths he introduced into the sciences of life.

Some thirty years ago, the medical school of Paris was still imbued with these doctrinal errors. I remember having been reprimanded at the Societé Philomathique, at the beginning of my career, by Professor Gerdy, who, invoking his surgical experience, expressed his opinion in the most categorical terms. "To say in physiology that vital phenomena are always identical under identical conditions is to promote an error," exclaimed Gerdy, "this is true only of inanimate objects."

---

*See my article in the *Revue Des Deux Mondes,* vol. IX, 1875.

The progress of the modern science of physiology and the more and more profound penetration of the physicochemical sciences into its culture have today almost dissipated most of these erroneous ideas, and it cannot be denied that present-day physiology follows a course which establishes more and more the rigorous determinism of the phenomena of life. It can be said that there is no longer any divergence among physiologists on this subject.

But it is not the same with the philosophers; they still reject physiological determinism and think that certain phenomena of life necessarily escape it; moral phenomena, for example. They fear that moral freedom could be compromised if absolute physiological determinism were accepted. Recently, even, a mathematician, on seeing the progress of this doctrine, sought to establish a reconcilation between scientific determinism and moral liberty.*

The misunderstanding between the philosophers and the physiologists comes without doubt from the fact that the word determinism is understood by them in the sense of *fatalism,* that is to say, in the sense of the philosophical determinism of Leibnitz.

The philosophers of whom we speak do not refuse to admit that the lower phenomena of animality might be subject to determinism, that movement and the functioning of organs might be regulated by it, but they exclude from this obligation the higher phenomena, the psychic phenomena, so that it would be necessary to distinguish in man the phenomena of life subject to determinism from those that are not.

For us, physiological determinism cannot be subject to restriction; all the phenomena that take place in living beings and in man, higher or lower phenomena, are subject to this law. "Every manifestation in the living being," we say, "is a physiological phenomenon and is associated with definite physicochemical conditions, which permit it when they are present and prevent it when they are absent."

Here is absolute determinism; it states that the psychic world is not outside the physicochemical world by any means, and this is a fact that experience verifies everyday. To be manifest, the

*Boussinesq, *Compt. rend de l'Académie. Revue Scientifique,* vol. XIX, p. 986, 1877.

phenomena of the soul require precisely determined material conditions; this is why they always appear in the same way, according to *laws,* and not arbitrarily or capriciously and by chance from a spontaneity without rules.

No one will contest that there is a determinism for moral *nonliberty;* certain alterations of the cerebral organ lead to insanity, cause moral liberty as well as intelligence to disappear, and becloud the conscience in the insane.

Since there is a determinism for moral nonliberty, there is necessarily a determinism for moral *liberty;* that is to say, an ensemble of anatomical and physicochemical conditions which permit it to exist. We affirm this fact and say that the manifestations of the soul, far from escaping physicochemical determinism, find themselves narrowly subject to it and never escape it, whatever the appearances to the contrary. Determinism, in a word, far from being the negation of moral liberty is on the contrary its necessary condition, as of all other vital manifestations.*

What would the world be if this were not so! The relations of what we call the physical with the moral would not be subject to the rule of precise laws, but would be in a state of anarchic discord or caprice, in a state contrary to the harmony of nature, without truth and without grandeur.

Determinism is thus nothing but the affirmation of *law* everywhere, always, and even into the relations of the physical with

---

*Freedom cannot be indeterminism. In the doctrine of physiological determinism, man is *forcibly* free; this is what can be foreseen; I do not wish to discuss the philosophical question here. It will suffice for me to say, from the physiological point of view, that the phenomenon of moral liberty ought to be equated with all the other phenomena of the living organism. If all the normal anatomical and physicochemical conditions exist in the arm, for example, and in the corresponding nervous organs, you can predict that you can move the member and that you can make it move freely in all directions according to your will. But the direction in which you will make it move exists as a future contingency that you cannot foresee, but which you are free to determine later according to the circumstances. In the same way, the anatomical and physicochemical integrity presumed for the cerebral organ makes you predict that its functions will be exerted fully and that you will be free to act voluntarily, but you cannot foresee the direction in which your volition will be exerted because this direction is, I repeat, controlled by the contingency of events of which you are ignorant and which you cannot foresee. This is why you remain free to act or to choose according to moral principles or any others that motivate you.

the moral; it is the affirmation according to the expression known from antiquity that, "Everything is done with order, weight, and measure."

The *law* of physiological determinism cannot interfere with moral liberty while, on the contrary, fatalism, that is to say, philosophical determinism, is in conflict with it and denies it.

In résumé, we shall propose the universality of the principle of physiological determinism in the living organism and shall express our thoughts as follows:

1. There are definite material *conditions* which regulate the appearance of the phenomena of life.
2. There are preestablished *laws* which regulate their order and their form.

CONCLUSION

The task that we proposed for ourselves in developing the considerations embodied in the three parts of this lecture was to eliminate from physiology certain problems that had been wrongly associated with it, various questions that are alien to it, and thereby to fix its scope and its purpose.

In the first part we have shown that in physiology it is necessary to renounce the illusion of a definition of life. We can only characterize its phenomena.

It is, moreover, the same in all science. Definitions are illusory; the conditions of things are all that we can know. In no order of science can we go beyond this limit, and it is a pure illusion to imagine that we can exceed it and that we can grasp the essence of any phenomenon, whatever it might be.

In the second part we have shown that materialist or spiritualist hypotheses are related to the search for primary causes that science cannot attain. In rejecting the search for primary causes we have thereby rejected the materialist hypothesis and the spiritualist hypothesis from the field of physiology.

In the third part we have accepted determinism as a necessary principle of physiology. Determinism reveals the conditions by which we can seize upon phenomena, suppress them, produce them, or modify them. This principle suffices for the ambitions of science because basically it reveals the *relations between phenomena and their conditions,* that is to say, the only and the true, *real, immediate,* and accessible *causality.*

We have thus disposed of the objection that is raised against physiologists of not knowing what *life* is. No one has made more progress elsewhere. Life is neither more nor less obscure than any of the other primary causes.

By saying that one should study nothing but the conditions of life, we circumscribe the field of physiological science, we fix the goal that we assign to it of conquering and mastering living nature.

Finally, in characterizing *life* and *death* by the two great types of phenomena of *organic creation* and of *organic destruction,* we embrace the whole of the conditions of existence of all living beings and we outline the program of studies which will be the object of the lectures that will follow.

SECOND LECTURE

The Three Forms of Life

SUMMARY: Life cannot be explained by an internal principle of action; it is the result of a conflict between the organism and the ambient physicochemical conditions. This conflict is by no means a discord but a harmony—Life presents three aspects to us which prove the necessity for physicochemical conditions for the manifestation of life—These three stages of life are: (1) Life in a stage of nonmanifestation or latency; (2) Life in the state of variable and dependent manifestation; (3) Life in the state of free and independent manifestation.

I. Latent Life—The organism falls into a state of chemical indifference—Examples taken from the plant kingdom and from the animal kingdom—Latent life is an arrested and not a diminished life—Conditions for the return of latent life to manifest life—Extrinsic conditions: water, air (oxygen), heat; intrinsic: reserves of nutrient materials—Experiments on the influence of air (oxygen)—Experiments on the influence of heat—Experiments on the influence of water—Phenomena of latent life in animals: infusoria, keronas, colpodas, tardigrades, anguillules of rusty wheat—Identification of the seed and the egg is not exact from the standpoint of latent life—Existence of beings in the state of latent life: brewer's yeast, anguillules, tardigrades, etc. —Explanation of the return from latent life to manifest life—Experiments by Chevreul on the desiccation of tissue—Mechanism of the transition to latent life—Mechanism of the return to manifest life—Necessary succession of phenomena of organic destruction and creation.

II. Oscillating Life—Belongs to all plants and to a great number of animals—The egg shows dormant life—Mechanism of vital dormancy—Influence of the external environment on the intern-

AL ENVIRONMENT—DIMINUTION IN THE CHEMICAL PHENOMENA DURING DORMANT LIFE—MECHANISM OF VITAL OSCILLATION IN DORMANCY—NECESSITY OF RESERVES FOR DORMANT LIFE—MECHANISM OF VITAL OSCILLATION—CESSATION OF DORMANT LIFE —INFLUENCE OF HEAT; IT CAN BRING ON DORMANCY LIKE COLD—RESISTANCE OF DORMANT BEINGS—ANIMALS AWAKENED DURING DORMANCY USE UP THEIR RESERVES RAPIDLY AND DIE—PHENOMENA OF CREATION AND DESTRUCTION IN DORMANCY—TEMPORARY DORMANCY DOES NOT REQUIRE RESERVES LIKE PROLONGED DORMANCY.

III. CONSTANT OR FREE LIFE—IT IS THE RESULT OF AN ORGANIC PERFECTION —OUR DISTINCTION BETWEEN INTERNAL ENVIRONMENT AND EXTERNAL ENVIRONMENT—INDEPENDENCE OF THE TWO ENVIRONMENTS IN ANIMALS WITH CONSTANT LIFE—THE IMPROVEMENT OF THE ORGANISM IN ANIMALS WITH CONSTANT LIFE CONSISTS IN MAINTAINING THE INTRINSIC AND EXTRINSIC CONDITIONS NECESSARY FOR THE LIFE OF THE ELEMENTS WITHIN THE INTERNAL ENVIRONMENT—WATER—ANIMAL HEAT—RESPIRATION—OXYGEN—RESERVES FOR NUTRITION—IT IS THE NERVOUS SYSTEM THAT IS RESPONSIBLE FOR THE EQUILIBRATION OF ALL THE CONDITIONS OF THE INTERNAL ENVIRONMENT—CONCLUSIONS RELATING TO THE INTERPRETATION OF THE THREE FORMS OF LIFE—ONE CANNOT FIND A VITAL, INDEPENDENT FORCE, OR PRINCIPLE. THERE IS ONLY ONE VITAL CONFLICT, WHOSE CONDITIONS WE MUST SEEK TO UNDERSTAND.

WE HAVE SAID that life cannot be explained, as it had been believed, by the existence of an internal principle of action acting independently of physicochemical forces, and above all, contrary to them. Life is a conflict. Its manifestations result from the intervention of two factors:

1. *Preestablished laws* that regulate the phenomena in their succession, their concert, and their harmony.

2. Definite *physicochemical conditions* which are necessary for the appearance of phenomena.

We have no influence on these laws; they are the result of what can be called the *anterior state:* they derive by atavism from organisms which the living being continues and repeats, and they can be followed back to the very origin of living beings. This is why certain philosophers and physiologists have believed they could say that life is only a *memory;* for myself, I have written that the seed seems to retain the memory of the organism from which it proceeds.

Only the conditions of vital manifestations are accessible to us. Knowledge of the external conditions that determine the appearance of vital phenomena suffice, as we have already said, for the purposes of physiological science, since it gives us the means of influencing and of mastering these phenomena.

For us, in a word, life results from a conflict, from a close and harmonious relation between the external conditions and the preestablished constitution of the organism. It is not by warfare against the cosmic conditions that the organism develops and maintains itself, but on the contrary, by an adaptation, an accord with these.

Thus the living animal does not constitute an exception to the grand natural harmony which makes things adapt to one another; it does not break any accord; it is neither in contradiction nor in strife with the general cosmic forces; indeed, far from that, it participates in the universal concert of things, and the life of the animals, for example, is only a fragment of the total life of the universe.

The nature of the relations between the living being and the ambient cosmic conditions permits us to consider three forms of life according to whether it is in a completely narrow dependence upon external conditions, or within a lesser degree of dependence, or in a relative independence. These three forms of life are:

1. *Latent life;* nonmanifest life.
2. *Oscillating life;* life with variable manifestations, and dependent upon the external environment.
3. *Constant life;* life with free manifestations, and independent of the external environment.

I. LATENT LIFE

Latent life, according to us, is exhibited by beings whose organism has fallen into the state of *chemical indifference.*

Tiedemann, as we have seen above, believed that life derived from an internal principle of action which prevented the organism from ever falling into the state of chemical indifference, so that the course of its vital manifestations could never be arrested or interrupted.

Observation and experiment do not permit the adoption of this proposition. We see beings which in some way live only potentially, without any manifestations characteristic of life. These beings are encountered in both the animal kingdom and the plant kingdom.

Active or manifest life, however attenuated it might be, is characterized by relationships between the living being and the environment; relations of exchange such that, at each instant, the being borrows liquid or gaseous materials from the cosmic environment and returns them to it. What characterizes the state of chemical indifference is the suppression of this exchange, the break in the relationship between the being and the environment which then remain face to face with one another, inalterable and unaltered. Thus a piece of marble, in ordinary conditions for example, remains in the atmosphere without appreciable change; it receives no action from it and exerts no action on it that is capable of modifying its chemical constitution.

Is it possible for living beings to fall into this degree of absolute chemical indifference? Certain physiologists have been reluctant to believe it, but there are cases where experiment obliges us to accept it. In the plant kingdom, seeds, and in the animal kingdom, certain reviviscible animals, Anguillules, Tardigrades, Rotifers, show us this state of chemicovital indifference. We already know a sufficient number of cases of latent life among the animals and the plants, but beyond these characteristic examples one can say without fear of error that latent life is profusely prevalent in nature and that in the future it will explain a great number of well-known facts reputed to be mysterious at the present time.

Seeds exhibit this phenomena of latent life to us. If all do not behave in an identical manner one can understand why and by what conditions latent life is less easily maintained in some than in others. It is in consequence of a greater or lesser alterability of their constituent materials by atmospheric agents.

It can be said that the life of a seed in the latent state is purely potential: it exists ready to manifest itself if it is furnished with the appropriate external conditions, but it does not manifest itself in any way if these conditions are absent. The seed has in itself, within its organization, all that it requires for life, but it does not live because it lacks the necessary physicochemical conditions.

It would be wrong to think that in this case the seed presents a life so attenuated that its manifestations escape observation because of the very degree to which they are weakened. This is not true, either in principle or in fact.

In principle, we know that life results from the concurrence of two factors, the one extrinsic, taken from the cosmic world; the other, intrinsic, derived from the organized body. It is a collaboration that is impossible to disjoin, and we ought to understand that in the absence of one of these factors, the being cannot live. It does not live when the conditions of the environment *do not exist* anymore than when they *exist alone*. Heat, humidity, and air are not life: organization alone does not constitute it any better.

In fact, we see seeds that have been stored for many years and centuries, which after this long inactivity can germinate and produce a new growth. These seeds have remained, during all this long period, as inert as if they were definitely dead. However attenuated the vital manifestations might be, the accumulation and prolongation of the exchanges would increase them somehow and make them appreciable. This reduced life ought to consume itself, yet under proper circumstances it does not do so.

Thus the seed contains within itself, within its intimate organization, all that it requires for life, but to bring this about it requires in addition the participation of external circumstances.

These circumstances are four in number.

Three *extrinsic* conditions:
- Air (oxygen)
- Heat
- Humidity

One *intrinsic* condition:
- The nutritive reserve of the seed itself.

This reserve is constituted by chemical materials which enter into the constitution of the seed and which act as a reservoir of food material which the vital manifestations will later expend.

But this is not all. It is also necessary for these conditions to exist to a certain degree, in a determined quantity; then life will shine forth in all its brilliance; beyond these limits life tends to disappear and as one approaches these limits the brilliance of the vital manifestations fades and diminishes.

A. Experiments on the latent life of seeds

We shall let you witness some experiments which are well known but which have a special interest here. Their object is to demonstrate that one cannot accept a free vital principle in these living beings when all the vital manifestations are closely related to physicochemical conditions whose enumeration follows:

1. Water

In some dry earth we have placed seeds that have also been dried, and which are at a temperature and in an atmosphere appropriate for vegetation. They lack only one condition, moisture; they are therefore inert. The wheat stored in Egyptian tombs called *mummy wheat* was, one might say, in the same situation. If they were provided with the humidity that they lacked, germination soon took place. I have consulted in this regard my savant colleague Mr. Decaisne, professor of agriculture at the Muséum. He told me that he considered all the examples of germination of seeds found in the hypogees to be false, because most ordinarily (as I was able to convince myself from a sample) these seeds were impregnated with bitumen or were carbonized. Germination of specimens coming from the lake dwellings would be equally most uncertain.

Even if one must discard these poorly observed facts from science, it has been established experimentally that seeds can germinate after more than a century. Among these seeds, one should cite those of the bean, of tobacco, of the poppy, etc.

It is necessary moreover that humidity does not prevent the access of air. Submerged seeds do not germinate, either because the dissolved oxygen is soon used up by the seeds or because it does not act in an appropriate state, that is to say, free. Nevertheless, submersion does not destroy the germinative faculty; there

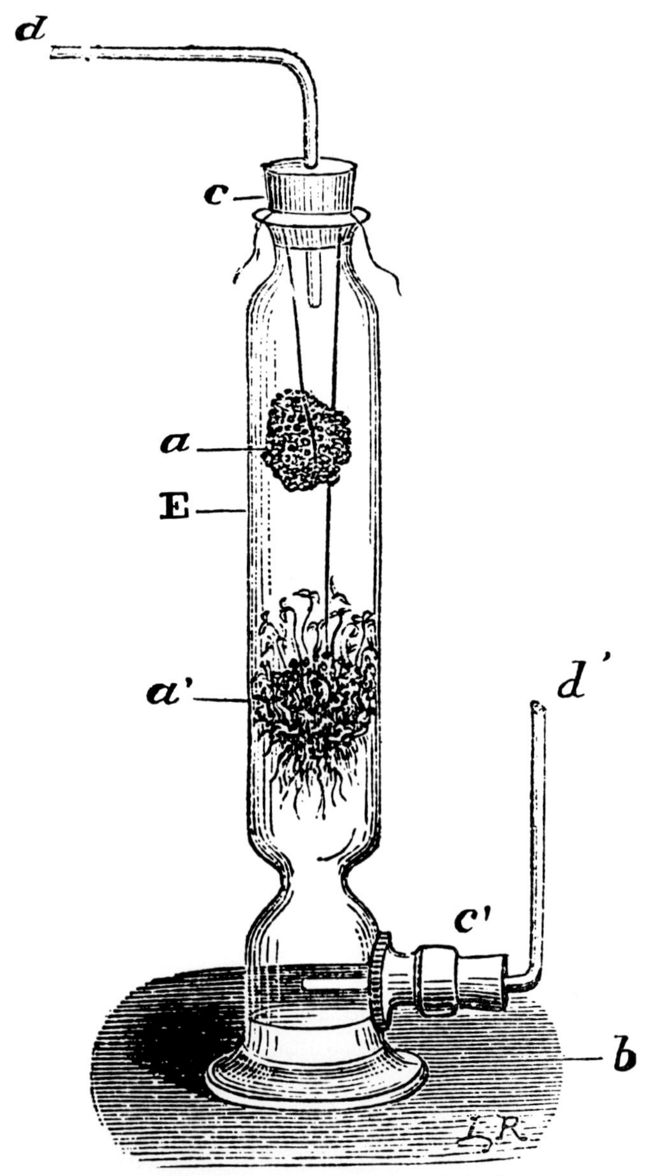

Figure 1. In this test tube E, we have introduced via the upper opening two moist sponges *a* and *a'* which are suspended by threads held by the rubber stopper *c*. The sponge *a* with seeds of garden cress just introduced into the apparatus; sponge *a'* with seeds of garden cress in the 4th and 5th day of

are even, according to Martins, seeds that can cross the seas from one continent to another and go on to germinate.

The simple apparatus that we shall use to germinate plants consists of a test tube (Fig. 1) in which we suspend with a thread wet sponges to which adhere the seeds that one wishes to germinate. At the bottom of the test tube we place a little water in *b* so that the sponge does not become dry; then the tubes *d* and *d'* are plugged or not, according to the desired circumstances, whether one wishes to confine the atmosphere of the test tube or circulate a current of air through it.

2. *Oxygen*

Here are test tubes in which seeds were placed on sponges at an appropriate humidity and temperature but in an atmosphere unsuitable for development. In one there is an atmosphere of nitrogen, in the other an atmosphere of carbonic acid.

We have chosen for these experiments the seeds of the common garden cress which have the advantage of germinating very rapidly. On a humid sponge in a test tube closed and filled with nitrogen, we have seen the seed swell; they have surrounded themselves with a sort of mucillagenous coating. The ambient temperature, from 21 to 25 degrees, was very favorable to germination, and nevertheless germination did not take place during the two or three days since the experiment began.

In another test tube we have placed seeds of garden cress in the same way on a humid sponge, in an atmosphere of carbonic acid, and germination did not take place either.

Finally, in a third test tube we have placed the seeds of garden cress in the same way in a moist atmosphere with ordinary air, and germination is already very evident after one day.

However, the seeds that have not yet germinated in the

---

germination. Two rubber stoppers *c* and *c'* are pierced by two tubes *d* and *d'* which permit the communication of the interior atmosphere of the apparatus with the external atmosphere. This makes it possible to pass different gases through the apparatus if desired or even to extract the gases it contains for analysis. In the bottom of the test tube there is a layer of water *b* so the internal atmosphere will always remain saturated with humidity.

atmosphere of nitrogen and of carbonic acid are not dead; the germination was only suspended, for if we make these gases disappear by substituting ordinary air or oxygen vegetation will soon be resumed.

These experiments show that to manifest its vitality the seed has need of all the conditions that we have enumerated above; if only one of them is lacking, water or oxygen for example, germination does not take place.

But this air itself must contain the proper proportion of oxygen. If it has too little, germination does not manifest itself; likewise, if it has too much, either when the atmosphere possesses a percentage composition too rich in oxygen, or when with its ordinary composition this air is compressed. Then, in a given volume, the proportion of the vital gas becomes too high, as the research of Bert has demonstrated.

In addition we have observed an important fact to which we shall return later. The seeds of garden cress, for example, cannot germinate except in an air that is relatively rich in oxygen; by mixing one volume of air with two volumes of an inert gas, hydrogen for example, germination does not take place. Strangely enough, all of the oxygen is absorbed. It seems probable that if then one were to add a new dose of oxygen to that which was insufficient at first to produce germination, it would be sufficient the second time. Respiration of the seed is thus very active and it appears up to a certain point to be relatively more intense than that of animals.

This necessity for an air rich enough in oxygen to promote germination explains to us how it happens that seeds that have been buried in the earth for a long time remain in a state of latent life and begin to germinate when they are returned to the surface of the soil. After deep terracing one has often seen the appearance of new vegetation which could only be explained in this way. I have it from an engineer that in certain terracings executed at the time of the construction of the Chemin de Fer du Nord, there appeared on the embankment a rich vegetation of white mustard which had not been observed before. It is probable that the changes in the terrain had exposed to the air the white mustard seeds buried in the soil and remaining in a state of latency at a depth which did not permit vegetation to take place because of the lack of oxygen.

3. Heat

Temperature must be kept within determined limits, but these limits are different for the various kinds of seeds. de Candolle has published some very interesting investigations on this subject in the *Bibliothèque Universelle and Revue Suisse* (Nov., 1865; Aug. and Sept., 1875). The fact that interests us here is the demonstration that for the same kind of seeds germination can be slowed or suspended, not only by too low a temperature, but also by too high a temperature. With the seeds of garden cress that have served in our experiments the temperature that seems to be most favorable for a rapid germination ranges between 19 and 29 degrees; above that, development appears to be difficult.

1ST EXPERIMENT. In test tubes arranged as described (see Fig. 1) we placed some days ago some cress seeds at the ambient temperature of the month of June, varying from 18 to 25 degrees. The very next day, after twenty-four hours, germination was very evident, the radicles had all sprouted and the folioles were beginning to disengage.

2ND EXPERIMENT. In four test tubes arranged as above we introduced the seeds of garden cress on damp sponges. We modified the experiment so that in the four test tubes we had a confined atmosphere. Instead of leaving the tubes *d* and *d'* open, we closed them by adapting to each of them a rubber tube which was compressed with a clamp.

Two of these test tubes were left in the ambient air of the laboratory (17 to 21 degrees). The other two test tubes were immersed in a water bath heated to between 38 and 39 degrees. By the next day the seeds had germinated in the two test tubes left in the laboratory, while no development had taken place in the test tubes sunk in the water bath. The third day germination was complete in the test tubes in the laboratory and those immersed in the water bath were as on the first day without any indication of germination. Then I removed one of the two test tubes from the water bath and placed it on the table beside those whose seeds were in full vegetation. The next day one could see no clear indication of germination but on the second and third day germination manifested itself and progressed actively thereafter. As for the other test tube remaining in the water bath at 38 to 39 degrees, on the seventh day it still showed no trace of germination; the seeds were spoiled, and surrounded by mold. This test tube was taken out of the bath and was placed on the table beside the others. Germination manifested itself, but very slowly; it did not begin to be evident until the third or fourth day. In the other experiments in which I left the test tubes more than eight days at a temperature of 38 to 39 degrees, germination never took place. Thus I have reason to believe that under the conditions as described, this point marks the upper limit for germination.

3RD EXPERIMENT. I placed other test tubes containing seeds of garden cress in a drying oven at 32°. They germinated very well, although perhaps a little slowly. Then I raised the temperature of the oven to 34.5°; then an arrest of germination took place. Sometimes two or three grew well nevertheless, but most often none germinated. I left seeds in the oven in this way for six to seven days without result. When they were taken out, the very next day germination was proceeding actively.

In résumé, it is seen that at 35 to 40 degrees, germination of garden cress is slowed or suspended, but not destroyed beyond recall. There is thus a sort of anesthesia, or rather of dormancy, produced by an excessive temperature as well as by too low a temperature. Thus the manifestation of vital phenomena requires not only the participation of heat, but a fixed level of heat for each being.

I would compare these experiments with another singular fact that I have noticed for a long time, namely that one can anesthetize frogs at this same temperature of 38 degrees, which is nevertheless the temperature of normal life in mammals.

We ought to make a remark here: the seed cannot be compared physiologically with the egg, as has been done too often. We shall see later that the egg never falls into the state of latent life. The seed is not the ovule, the germ of the plant; it is its embryo. The essential part of the seed is in fact the miniature of the complete plant; in it are found the rudiment of the root, or *radicle,* the rudiment of the stem, or *tigelle,* the terminal bud, or *gemmule,* and the first leaves, or *cotyledons.*

It is therefore the *embryo* that remains in the state of latent life as long as the external conditions are not conducive to its development.

It results from this that what we have said previously about latent life does not apply to the plant egg but rather to the plant itself.

Water and heat are for the plant embryo the indispensable conditions for the return from latent life to manifest life. Suppression of these conditions always causes life to disappear, their return makes it reappear.

A curious experiment by Th. de Saussure shows that even when the embryo has begun its germinative development, it can still stop and fall back into chemical indifference. Germinating wheat is taken and dried; in this state it can be stored for a long time, absolutely inert, just as the seed from which this embryo arose is stored. The air confined in the vessel that contains the dried embryo undergoes no modifications, and gives evidence thereby that the exchange between the rudimentary being and the environment is nil. By restoring humidity and heat, that is to say, propitious conditions to it, life reappears. This can be repeated alternately a sufficiently great number of times and the result produced is always the same. The faculty of latent life will disappear only when the development is advanced enough for green matter to appear in the first leaves.

These phenomena of latent life explain certain very remarkable natural circumstances which forcefully struck the imagination of those who observed them for the first time.

A large number of true seeds or spores (simple seeds of the acotyledons) are buried in the soil or scattered on the surface in a state of inertia. Suddenly, following an abundant rain or the reworking of the soil, they enter into germination and the ground is covered with an unexpected and seemingly spontaneous vegetation.

Similarly, on garden paths after a thunderstorm, one sees

green plaques formed by the development of a species of algae, the Nostoc.

None of these growths appeared suddenly or spontaneously; the seeds existed in the depths of the soil or in a state of desiccation in the dust that covered it and they did not show any development until they found the conditions of aeration, of humidity, and of heat that are the three essential factors for vital manifestations.

B. Latent life among animals

Animals organisms also afford many examples of latent life. A large number of beings are susceptible to lapsing into a state of *chemical indifference* through desiccation. Such are many of the Infusoria, the Colpoda among others, which have been well studied by Coste, Balbiani, and Gerbe *(Compt. Rend. de l'Acad. Des Sci.*, vol. LIX, p. 14). But the best known of these animals are the *Rotifers*, the *Tardigrades*, and the *Anguillules of rusty wheat*.

The *Colpoda* are ciliated Infusoria of quite large size, having the form of a bean, armed with vibratile cilia over all their surface (see Fig. 2*e*). Under the microscope they are seen to take monads, bacteria, vibria, etc. into their stomach via a mouth placed in an indentation in their body, and expel the residue of their digestion by an anal aperture placed at the broad end of the body. Near this anal aperture is found a contractile vesicle taken for the heart by certain micrographers, which seems to be the propulsive organ of an aqueous apparatus. At the center of the body of the Colpoda appears a quite voluminous organ of reproduction.

When on the surface of infusions there forms a pellicle in which monads, vibria, and bacteria develop, the Colpoda dispersed throughout the vessel can be seen to direct themselves toward this pellicle to satisfy their hunger on the animalcules which compose it, or instead, to place themselves in contact with the air. Then, among these Colpoda, some are seen to stop suddenly, begin to rotate in place, roll up in a ball, and continue this gyration until a secretion from their body coagulates around them into an enveloping membrane; in short, they become encysted, and then

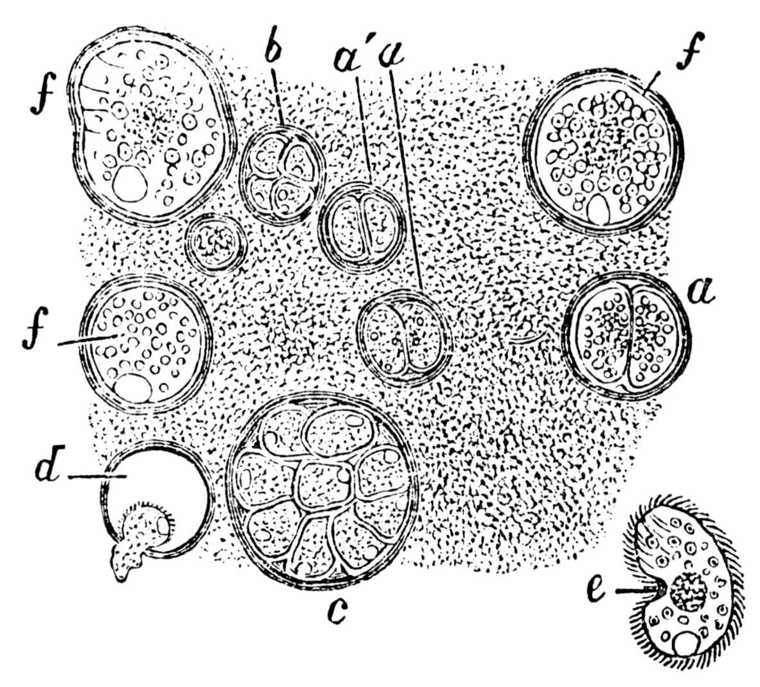

Figure 2.—Encystment of the Colpodas *a, b, c,* Colpodas dividing into two; four and greater numbers of new Colpodas within their cysts; *d,* Colpoda emerging from its cyst; *e* free Colpoda; *f, f,* encysted Colpodas.

they become completely immobile in their envelope like an insect in its cocoon. At this period of their existence, the smallest ones have a great similarity to an ovule; it is this which might have led to the belief in a *spontaneous* egg.

Soon the encysted and immobile Colpoda separate into two, four, and sometimes twelve smaller Colpoda (see Fig. 2) which, once separate and distinct, enter into a gyration on their own account within their common envelope. The movements to which they devote themselves finally wear through the cyst at some point, and once a fissure is made they can be seen leaving their prison and mingling with the population whose number they increase. These are *multiplication* cysts in contrast to another encystment which is related to the conservation of the individual. This is the explanation for multiplication in infusions.

When the Colpoda in infusion have exhausted their reproductive power and evaporation threatens to dry up their container, they encyst in order to protect themselves against destructive influences. They can then be dried on glass slides and preserved in this state indefinitely; they return to life as soon as moisture is restored to them. In this way Balbiani has preserved individuals for seven years, restoring them to active life and drying them up each year.

The Colpoda cysts, impalpable animal seeds, attach themselves like dust to the surface of objects, on leaves, branches, the bark of trees, on the grass at the bottom of dried up ponds, in the sand or in the dried up mud. Their small size permits them to pass through filters, and one cannot get rid of them. They rupture their envelope each time rain or dew restore humidity to them, take the food within their reach, and form a new cocoon as soon as they begin to lack water. They therefore pass by turns through states of apparent death and resurrection, under the influence of a physical condition which is present or absent.

The *Rotifers,* or rotators (Figs. 3 and 4), are animals that are already at a high level of organization classified either among the worms (Gegenbaur) or as a separate group between the crustaceans and worms (Van Beneden).

These animals are from 0.05 to 1.0 mm in length; they are therefore far from being microscopic. They are found in mosses and especially in those *(Bryum)* which form in green turfs on roofs. Their organization exhibits quite a variety of apparatus to us; they possess quite complicated visceral and locomotor organs (see Fig. 3). They can creep or swim and according to whether they have recourse to one mode of locomotion or the other the aspect under which they present themselves changes. In the most ordinary state their body is fusiform, thinner at the anterior end and terminating in a sort of ciliated suction cup by means of which they attach themselves to solid bodies to progress by reptation like leeches. At other times, this prolongation is retracted to the inside and then two round lobes are seen to project in the form of discs edged by cilia. In the state of latent life they are immobile and gathered into balls as seen in Figure 4.

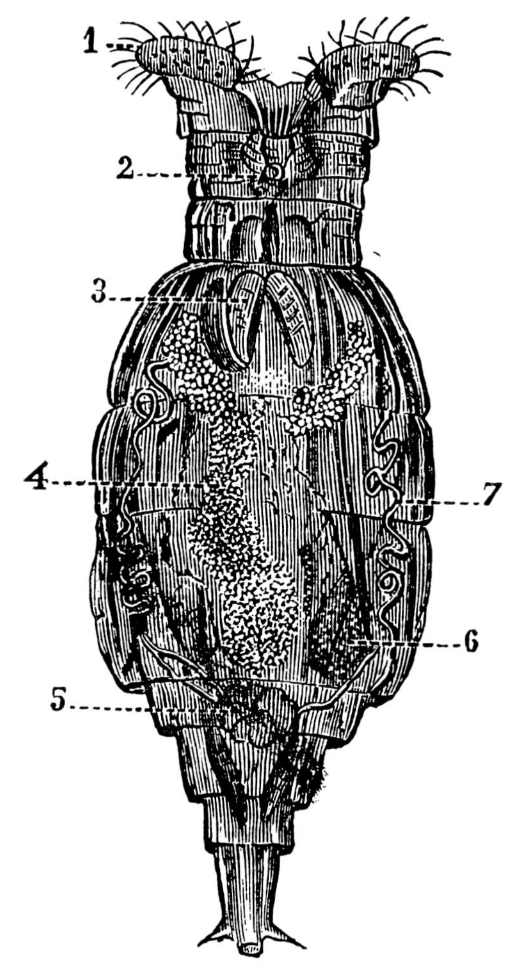

Figure 3.—Rotifer of roofs in the state of active life. 1, ciliated organs; 2, respiratory tube; 3, masticatory apparatus; 4, intestine; 5, contracile vacuole; 6, ovary; 7, excretory canal.

The *Tardigrades* (Fig. 5), well studied from the point of view of their latent life by Doyère,* are still more highly organized animals than the preceding. They belong to the class *Arachnida:* it is a family of the *Acarina.* They have four pairs of short legs,

*Doyère, *Ann. des. Sc. Nat.,* 1840-1841.

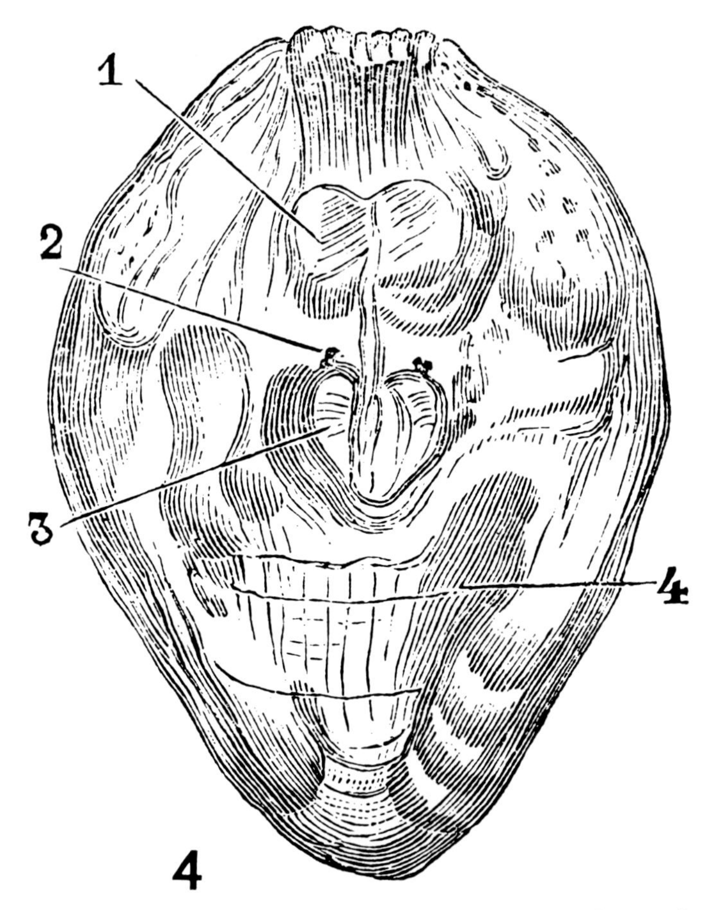

Figure 4.—Rotifer in the dessicated state. 1, rotatory organ; 2, eyes; 3, masticatory apparatus; 4, intestine.

articulated and provided with nails. Their body, pointed in front, permits one to distinguish three or four articulations.

Strictly walkers, these animals live in the dust on roofs, or on the mosses that grow there. Exposed to excessive hygrometric variations, they live sometimes in the water which bathes the sand of the gutters like true aquatic beings, sometimes like earthworms.

When they are short of water, they retract, shrivel up, and mix with the surrounding dust; they can remain several months, and

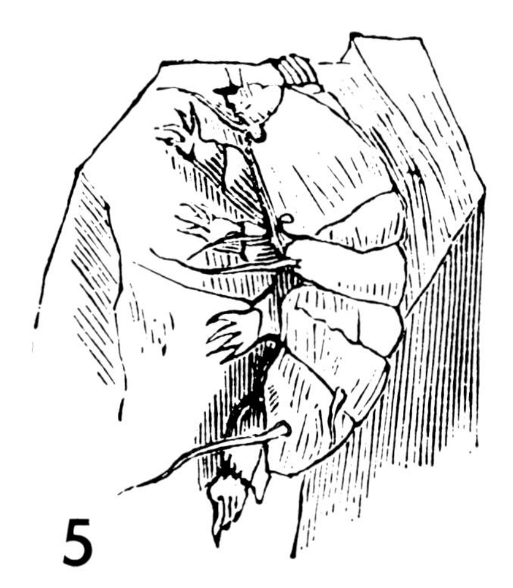

Figure 5. Sketch of the Tardigrade *(Emydium testudo)* climbing on a grain of sand.

it is conceivable that they can remain indefinitely in this state of desiccation without appreciable manifestations of life.

But when this dust is moistened, as Leeuwenhoek did for the first time on the 27th of September 1701, within an hour the animals can be seen swarming therein, active and mobile; their organs, muscles, nerves, digestive viscera reassume their forms (see Fig. 6 and 7) ; in a word they resume the fullness of their vitality until desiccation comes to interrupt it again.

These facts have had wide reverberations and at other times given rise to discussions about the question of knowing whether in truth life had been suspended completely during desiccation or only attenuated as is produced by cold in hibernating animals. After a debate brought before the Société de Biologie by Doyère, Davaine, and Pouchet, it was firmly established that: "(1) there is no appreciable life in the inert bodies of reviviscible animals, and (2) these bodies retain their property of revival in conditions (vacuum, dry, at 100°) *incompatible with any kind of manifest life."*

According to these facts, it seems quite certain that life is completely arrested despite the complexity of the organization

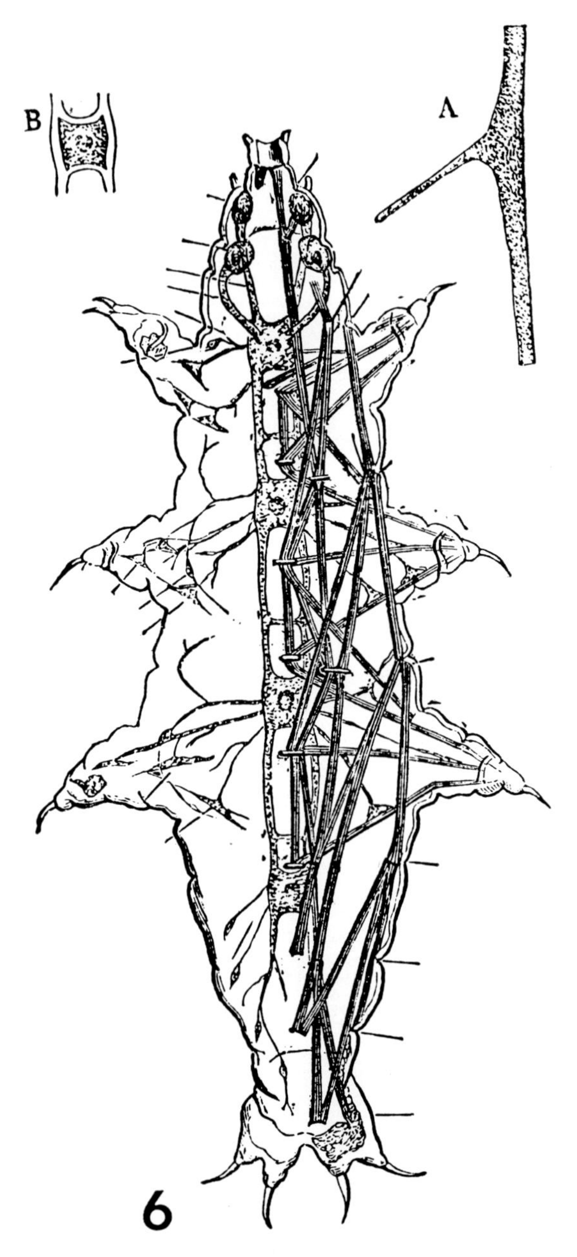

Figure 6. Muscular and nervous system of a *Milnesium tardigradum* (figure taken from Doyère, Thèse de la Faculté des Sciences, Paris, 1842).
Muscular and nervous systems of the Tardigrade—A, mode of termination of the nerves in the muscles; B, a nervous ganglion of the subintestinal chain.

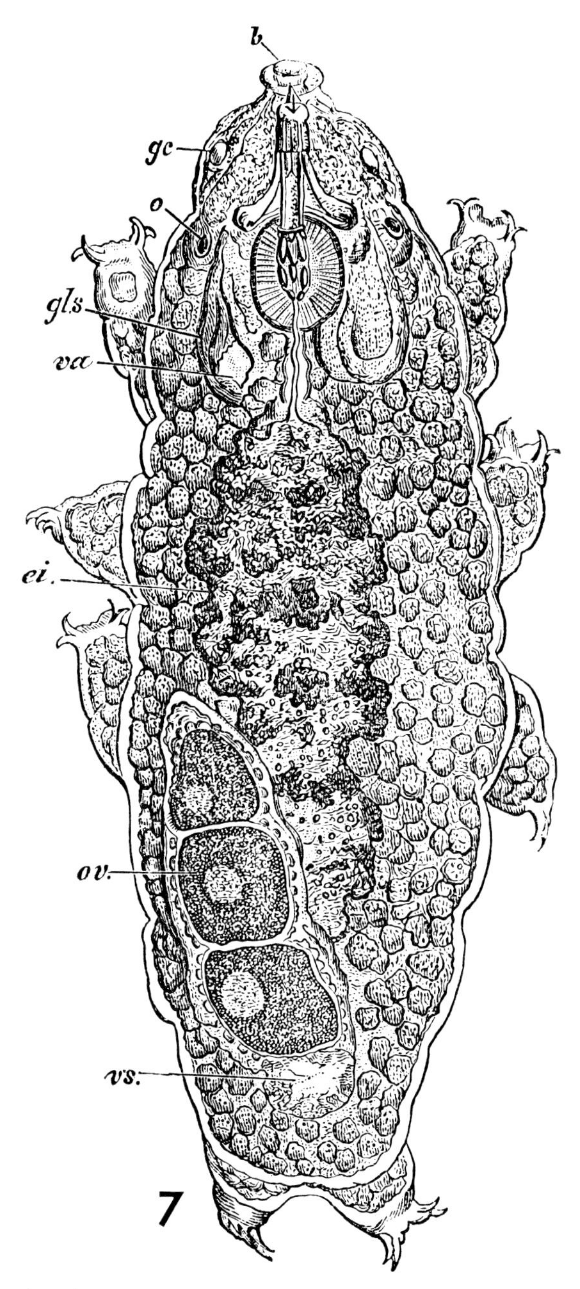

Figure 7. Digestive system of *Milnesium tartigradum* (Doyère, Thèse de la Faculté des Sciences, Paris, 1842).
b, mouth; *gls,* salivary glands; *ei,* digestive sac with its exterior lobes and internal cavity; *ov,* the ovary displaced to the side; *vs,* seminal vesicle.

of these animals. In them in fact are to be found muscles, nerves, nervous ganglia, glands, eggs, and in a word, all the tissues that constitute higher organisms (see Fig. 6 and 7). To my knowledge, however, no one has ever carried out the experiment of storing them for a very long period of time in the state of latent life. The true *criterion* which permits decision whether life is really arrested in an absolute manner is the indefinite duration of this arrest.

Anguillules of rusty wheat (Fig. 8). The facts observed regarding the anguillules of rusty wheat are no less interesting than those we have examined above. They lead moreover to the same conclusions.*

Rust manifests itself in wheat by a deformation of the grain after maturity and by a change in color. The grains are small, rounded, and blackish, and consist of a thick hard shell whose cavity is filled with a white powder (Fig. 8A and B). This disease is produced by the presence of very small nematode helminths, present in each seed in many thousands. These worms *(Anguillula tritici)* have no sexual organs at all and cannot reproduce: but they come from eggs laid by other worms provided with genital organs which had penetrated the seed before its maturity. They introduced themselves into the developing young plant, during germination, between the sheaths of leaves which enclose the head in the process of formation (Fig. 8C).

But this introduction is only possible if the plant is moist, because only then is the worm active and able to climb the length of the stalk. If not, the worm will remain in the soil at the foot of the new head, and the wheat will be protected from its attack. Thus it is in the wet years when the rains are abundant during the time of formation of the head that the grain is subject to rust. The farmers knew that, but they could not understand the relationship between the wetness of the season and the rust of the wheat. It can be seen that this relationship has nothing mysterious about it; it is a simple physical condition which determines whether the pathway is accessible to the parasite or not. It is generally so, and all the natural harmonies derive from physicochemical conditions when we know their mechanism.

*Davaine, *Mémoires de la Société de Biologie,* 1856.

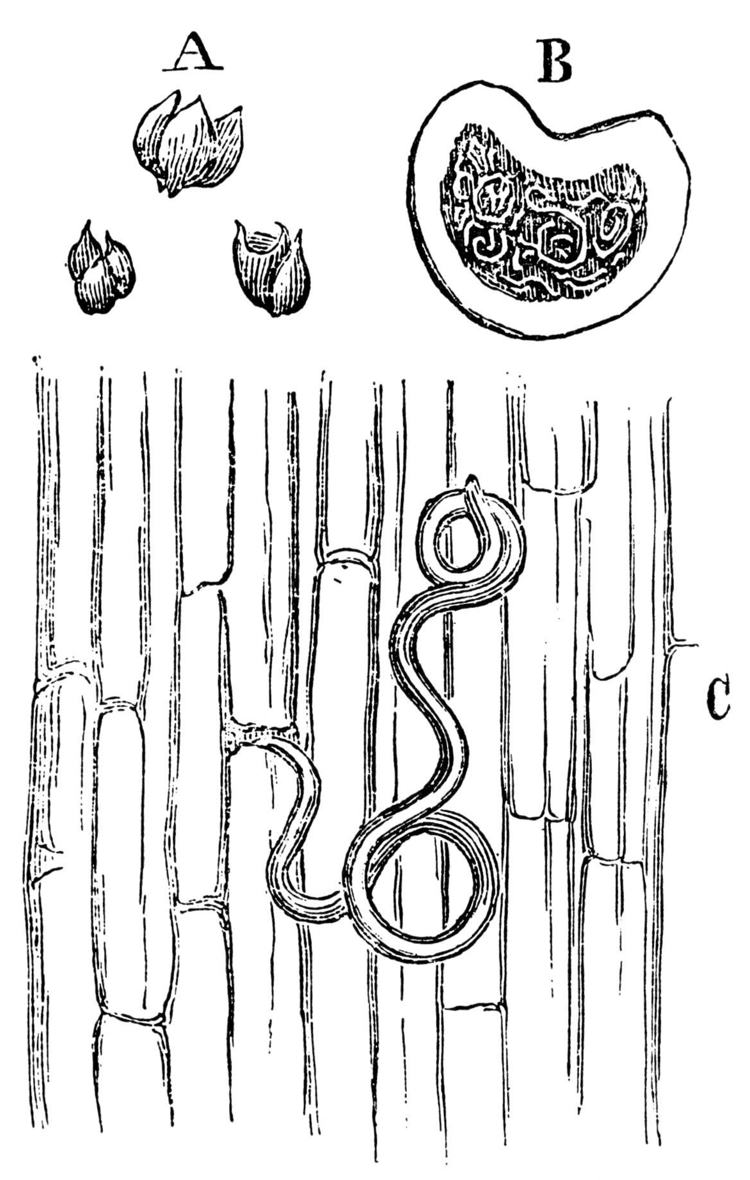

Figure 8.—Figure according to Dr. Davaine *(Mémoires de la Société de Biologie,* 1856)
A, grains of rusty wheat in their natural size.
B, cross-section of rusty grain containing adult Anguillules magnified four times.
C, longitudinal section of a young stalk of wheat, enlarged a hundred times; it was not possible to depict more than a portion of this section on which one sees a larval Anguillule, its position shows that it is neither within the vessels nor in the tissue of the leaf, but on the surface.

At this time the grain of wheat is composed of a young and soft parenchyma in which the various parts, the paleola, stamens, and ovaries are not at all distinct and into which the worm can easily penetrate. It is there that the animal passes from the larval to the adult stage; its sexual organs, which had not yet developed, appear and attain their organic perfection; the female lays eggs which hatch and live in a larval state in the cavity which shelters the parents, doomed to perish. The larval worms soon dry up with the wheat itself and await, in a state of apparent death, the conditions necessary for their vital manifestations, humidity and air.

The larval worms present themselves in the form of a white powder, grossly resembling starch, having an average length of eight tenths of a millimeter (Fig. 8B).

The respiration of these animals when they are in the grain of wheat is nil. Davaine has maintained worms enclosed in green heads in vacuum for twenty-seven hours, without the activity of these animals being modified noticeably by this treatment. It can thus be conceived that it would be possible to preserve desiccated worms indefinitely in a vacuum. But living larvae in water cannot be treated in the same way. Exposed to a vacuum, they soon fall into a state of apparent death; they return to activity when the air is allowed to return again. I have shown you that it is enough to prevent the contact of air with the water in which they live, by placing oil, for example, around the cover slip of the microscope's slide, to see the worms soon fall into a state of asphyxia.

Davaine, having found in the intestines of these animals neither a cellular lining to which digestive functions could be attributed, nor solid particles, concluded that apparently the nutrition of these animals, like their respiration, is accomplished in part through the skin. I think that nutrition must above all take place by means of alimentary reserves which the body of the animal contains and not by absorption of substances coming from outside.

These animals move in a fixed position without really making any progress as long as their life lasts. Their movements are not subject to interruption unless some external condition intervenes.

Desiccation or withdrawal of air are the ordinary conditions that stop these movements as well as all the apparent manifestations of life.

In 1771 Baker observed that worms kept inert for twenty-seven years resumed their activity as soon as they were moistened. For my part I have seen worms come back to life after having been stored for four years in a very dry and well-stoppered flask.

Spallanzani was able to induce their revival and their dormancy as many as sixteen times in succession. These animals cannot return to life indefinitely because with each revival they consume a part of their nutritive materials without being able to restore this loss, since they do not eat. Thus in the end, the intrinsic condition constituted by the reserve of nutrient materials eventually disappears and prevents life from manifesting itself even though the three other extrinsic conditions, heat, water, and air, remain.

If the temperature of the water containing the worms is lowered progressively, they retain their movements as far as zero. Then the movements die out. Then, when the temperature is again increased, it is only at about 20 degrees that they can be seen to leave their state of apparent death. They revive even when they have been subject to a considerable drop in temperature, to about 15 or 20 degrees below zero. They are much less resistant than rotifers to high temperature, and at 70 degrees above zero they inevitably perish.

It has been observed that it is necessary to continue the action of moisture for quite unequal periods of time to produce revival in the worms. But it can be arranged so that only one of the other necessary conditions is lacking, aeration for example; when this is made to intervene after prolonged humidification, revival will take place in nearly equal periods of time. To carry out the experiment I moistened rusty grain of wheat for twenty-four hours; then on opening them it was seen that just about the same time is necessary to restore the animals to the possession of their vital functions. At any rate, if whole rusty grains of wheat are left immersed in water for entirely too long a time, the worms eventually lose the faculty of reviving.

Other Examples of Latent Life: Eggs, Ferments, Brewer's Yeast, etc.

We have seen that the seed provides one of the clearest examples of latent life. The substratum of life is indeed present in the seed, but when the external physicochemical conditions are lacking, all conflict, all vital motion is suspended.

An attempt has been made to look for analogous phenomena in the eggs of certain animals by comparing them to seeds. Such an assimilation is not correct. The seed is not an egg, as we have already said; it does not have the properties of one, it is an embryo.

It is not surprising moreover that the egg cannot like the seed fall into a state of chemical indifference, in the state of latent life. The egg is a body in the course of evolution, whose development cannot be arrested completely. It is only in the state of dormant or oscillating life; as we shall see, it always remains in a relationship of material exchange with the environment. In a word, the egg breathes; it takes in oxygen and gives off carbonic acid; it does not remain inert in an unaltered, ambient medium.

The indifference or apparent inertia of the egg is only an illusion produced by the slowness, the attenuation, or the obscurity of the phenomena that take place within it. The eggs of the silkworm, for example, await the return of spring to hatch; but it must be admitted that life therein has not been completely suspended. Changes take place in them under the influence of cold, and on the return of spring warmth does not find an egg in the same state and with the same constitution that it had at the end of autumn. From this it can be understood that the heat which previously could not cause the development of the egg, can now do so.

These phenomena resulting from the influence of the physical conditions of the environment on the latent or the dormant life of beings explains to us certain harmonic adaptations of nature. What would be the use, for example, for the silkworm egg to hatch in the middle of winter, since the animal would not find any of the leaves upon which it must nourish itself? It is thus natural for this egg to acquire this faculty only in springtime and for it to lie dormant during the cold of the winter while slowly completing its development. Phenomena similar to hibernation no doubt take place in plants. We should not however attribute these phenomena to supernatural or magical causes. The influence of the course of the seasons and the influence of their duration are explained by the repetitions and successions of definite physico-chemical conditions. Winter has not acted upon the eggs of the silkworm as a special or extraphysical condition; winter has simply acted as a physical condition, as *cold*. This is what the experiments of Duclaux have demonstrated. A silkworm egg laid at the end of the summer will not hatch naturally until the following spring because the winter and cold bring about a physical condition favourable to certain imperceptible developments which must precede its hatching. But natural winter can be replaced by an artificial winter. If these eggs are subjected for twenty-four hours to the action of a temperature of zero degrees, then upon the intervention of heat, development takes place immediately and without delay.

The *ferments,* those agents of life so important and still so little known have the faculty of falling into a state of latent life. However, we must make a distinction here in regard to *soluble* ferments and *formed* ferments. The first are not living beings, and the property that they exhibit of drying up, then redissolving and recovering their chemical activity does not resemble, except distantly, the phenomena of latent life. Formed ferments, on the contrary, are living beings which reproduce; after having been desiccated they revive under the influence of humidity, and manifest not only their chemical properties but also their property of proliferation, or of reproduction; these are indeed true phenomena of *latent life*.

Brewer's yeast provides us a good example of this double faculty. When yeast is taken in full activity and submitted to gradual desiccation, it will become reduced to the state of latent life, it can be exposed to a much higher temperature or to the prolonged action of alcohol and it will resist these trials; then, when placed in proper conditions it will revive and be able to develop again.

Here is a tube in which we set into fermentation some brewer's yeast that had been dried at 40 degrees and stored for two years; little by little it absorbed water and produced alcoholic fermentation when sugar was added.

In another tube we placed brewer's yeast equally desiccated and stored in pure alcohol for a year and a half. It also absorbed water little by little and successfully produced alcoholic fermentation.

In another experiment I suspended fresh brewer's yeast in absolute alcohol, where it remained immersed for three or four days. Following this, I collected the yeast on a filter to dry it; when again immersed in sugar solution, it gave rise to a very active alcoholic fermentation. I must add that in all cases where yeast has been dried beforehand, whether or not it has been subjected to the influence of alcohol, it is necessary for it to be soaked again in a preliminary maceration for twenty-four to thirty-six hours before alcoholic fermentation appears with all its characteristics of inversion of the sucrose into glucose, splitting of the glucose into carbonic acid and alcohol, etc. Thus, it is seen that the two ferments of which brewer's yeast is constituted, the inversive ferment or soluble ferment, and the *torula cerevisiae* or formed ferment, both possess the faculty of regaining their properties after desiccation.

Explanation of Latent Life

Desiccation is a protective state for organisms that must be exposed to atmospheric vicissitudes. We have seen Colpoda, Rotifers, Tardigrades, and Anguillulles encyst, divide, roll up, etc., as soon as the water necessary for their vital manifestations is lacking.

If now we seek to give an account of the mechanisms by which the state of latent life is produced and the return to manifest life is made, we shall see with the greatest clarity the influence of external conditions manifested upon the two orders of phenomena with which we have associated life in all the beings, *organic creation* and *destruction.*

Let us first take up the passage from manifest life to the state of latent life. The principal condition that an organism must fulfill to fall into this state is *desiccation*. The other circumstances, temperature, the composition of the atmospheric gases, cannot act as effectively as desiccation to suspend life. A moist seed submitted to cold or exposed to an inert gas would probably finally decay. However, it could not be concluded absolutely that the unlimited maintenance of latent life demands desiccation, because seeds buried in the earth or under water are preserved in the state of latent life for indeterminate but certainly very considerable lengths of time (at least a century).

The immediate consequence of desiccation is to cause to disappear and to render impossible the phenomena of *organic destruction,* that is to say the functional manifestations of living beings; it is equally true for other conditions that produce latent life. The physical properties of tissues, their elasticity, their density, their tensile strength, are first modified by an excessive degree of desiccation of the organized substance. Then follow the chemical phenomena of vital destruction, whose action is arrested by the very fact of desiccation, for the agents of these phenomena, the ferments, become inert when desiccated. Desiccation thus brings about the suppression of vital *destruction* by causing the physical and chemical properties of the tissues to disappear. Vital *creation* itself then also stops in the desiccated cells. In short, life considered in its two aspects is suspended; the organism is in a state of chemical indifference; it is inert. There is an arrest of life, or *latent life*.

The influence of desiccation on the *physical properties* of the tissues and the substances of the organism has been emphasized in a fundamental work published in 1819 by Chevreul *(Mémoires du Muséum,* vol. XIII).

These investigations, of great importance to physiology, were carried out on tendons, fibrous tissues, the ligamentum flavum, and various albuminoid substances.

The *tendons* comprise the tissues by which muscles are attached to the bones; they present themselves in the normal state as supple cords, elastic, of a pearly appearance, having great tensile strength. When they are dried, they lose about 50 per cent of water, they become yellowish, their elasticity is diminished to the point that if they are bent, tears and breaks are produced, and the tissue is disorganized. But then the tendon is replaced in the water, it absorbs this liquid again until it regains almost its normal content. Desiccation caused it to lose its properties; moisture restores them to it.

The *fibrin* of blood finds itself in the same circumstances. It can lose 80 per cent of water by desiccation, and with this its color, its tensile strength and its elasticity disappear. Replaced in contact with water, it regains about the same quantity and recovers its lost properties.

The transparent *cornea* exhibits similar phenomena. When dried, it becomes opaque; moistened again, it regains its transparency.*

It is thus seen that for tissues which can be considered as simple physical materials of the organization, their properties enter into the manifestations of life only by virtue of the water they contain.

The soluble *albumen* of the egg exhibits phenomena very similar to those which we have indicated above.

If it is desiccated slowly (below 45 degrees) it becomes yellow and brittle, losing about 90 per cent of water. If water is then added, it redissolves again. When the albumen is found in this state of desiccation, it can be subjected to a high dry temperature, to 100 degrees for example, without losing its ability to redissolve.

Egg albumen *coagulated* by heat can be desiccated by evaporation of about 90 per cent of water, but, if after desiccation it is

---

*It is not only desiccation that causes the cornea to lose its transparency. When the eye of a dog or of a rabbit recently removed from the socket is compressed between the fingers, the cornea is seen to become opaque under pressure and to regain its transparency when the compression ceases. Quite a long time ago I showed that this phenomenon could be reproduced during life. If the eyes of a dog or of a rabbit are made to extrude with the end of the handle of a scalpel, the two eye balls extrude with a cornea that is opaque to such a degree that the animal becomes blind, but as soon as the eye is pushed back into the socket, the compression having ceased, the cornea becomes transparent and the animal recovers its sight. Here the opacity of the cornea must not be attributed to desiccation of the cornea but rather to a change in the molecular disposition of its constituent parts.

moistened, it can be seen that it has lost forever its ability to redissolve. This experiment on the solubility of albumen in its various states is a salient fact from the point of view of the subject that concerns us.

We see how the suppression of humidity and of favorable extrinsic conditions can bring about the disappearance or at least the suspension of the properties of tissues; every vital manifestation that requires the participation of these physical and mechanical properties is itself suppressed in virtue of this.

We must relate to these facts an experiment by Glénard, of Lyon, concerning the desiccation of the blood of the horse within its vessels. Horse blood coagulates slowly; the blood contained within a jugular vein, for example, is desiccated at a temperature below 45 degrees. After desiccation, it can be ascertained that this blood will redissolve in water and that the plasma that results therefrom has not lost the property of coagulation. This demonstrates the interesting fact, that in a higher animal, just as in the lower beings, the soluble fibrin of the plasma does not lose its property of coagulation by desiccation.

We have said that desiccation, that is to say the disappearance of the humidity necessary for organisms, suppresses not only the physical properties of the tissues, but also the *chemical phenomena* that take place therein. We know that these phenomena have ferments as their principal agents, and that here it is a matter of fermentation. But the simplest of experiments show us that these fermentations, like all chemical actions, cannot take place except within a liquid medium. *Corpora non agunt nisi soluta.*

For fermentations to take place, therefore, an appropriate degree of both temperature and humidity is necessary, otherwise the action is suspended. For a long time I have demonstrated in my courses that ferments have the property of desiccating and regaining their properties when they are again moistened. Here is some pancreatic ferment in the dry state; it can be placed in contact with dry starch without producing any action at all. If water is added, the transformation into sugar will take place rapidly at the proper temperature. The ferment had not therefore lost the ability to act; it was only unable to manifest its action.

Desiccated gastric juice no longer digests; it can remain indefinitely in contact with meat that is also desiccated without attacking it. The addition of water at a temperature close to that of the body, at 40 degrees, will cause the suspended digestion to reappear.

It can be understood from these examples that desiccation abolishes the two orders of physical and chemical phenomena of the organism. Those phenomena characterizing vital destruction being prevented, organic creation stops in turn; the organism loses the characteristics of life.

The awakening of a being deep in the state of latent life, its return to manifest life, is just as simply explained.

First it is vital destruction that again becomes possible through the return of the physical and chemical phenomena: then creative life reappears in turn when the animal again takes in nourishment.

As soon as humidity and heat are restored to the organism, the tissues, as the investigations of Chevreul have shown, regain the quantity of water that they had before their desiccation, and their mechanical and physical properties of resistance, elasticity, transparency and fluidity reappear. The return of the chemical phenomena takes place just as quickly; the desiccated ferments, moistened again, regain their activity, the interrupted fermentations resume their course within the living organism as well as outside of it, as direct experiment has shown us.

It is therefore primarily by the reestablishment of the acts of vital destruction that the return to life takes place. Creative life evinces itself only as a second event. This is a law that should be emphasized.

In its rebirth the animal or the plant always begins by destroying its organism, by consuming the materials previously placed in reserve. This observation makes us understand the necessity for a new condition of reviviscence or return to manifest life. It is necessary for the being to possess reserves, accumulated in its tissues, to be able to nourish itself and provide for its initial expenditures, until the time when it is completely restored to life, and is able to draw from the outside, by alimentation, the materials that are necessary to make new reserves. Here again, incidentally, we encounter an application of that great law upon which we shall never cease to insist, namely, that nutrition is always indirect instead of being direct and immediate. The accumulation of reserves is therefore a necessity for beings in latent life; the resumption of vital manifestations is possible only at this price.

Once the phenomena of vital destruction have begun again within the being, which was inert only a short time ago, vital creation also resumes its course, and life is reestablished in its integrity with its two orders of characteristic phenomena.

II. OSCILLATING LIFE

The living being, considered as a complex individual, can be linked to the external environment in such a close dependency that its vital manifestations, without ever being completely extinguished as in the state of latent life, are nevertheless attenuated or enhanced in very large measure as external conditions vary.

The beings whose vital manifestations can vary within wide limits under the influence of cosmic conditions are beings with a life that is *oscillating* or *dependent* on the external environment.

These creatures are very numerous in nature.

All plants are of this nature: they are dormant during the winter. Life is not completely extinguished within them; the material exchanges of assimilation and dissimilation are not absolutely abolished, but they are reduced to a minimum. Vegetation is hardly visible; the vital processes are nearly imperceptible. In the spring, when warmth returns, the vital movement is enhanced, the dormant vegetation assumes its utmost activity, the sap sets in motion, leaves appear, buds open and develop, and new parts, roots, and branches spread out into the soil or the air.

Similar phenomena take place in the animal kingdom. All invertebrates, and among the vertebrates, all cold-blooded animals, possess an *oscillating* life, *dependent* on the cosmic environment. Cold makes them dormant, and if they cannot be removed from its influence during the winter, life becomes attenuated, respiration slows down, digestion is suspended, movements become feeble or disappear. In mammals this state is called the *state of hibernation;* the marmot and the dormouse serve as examples.

It is usually a lowering of the temperature that produces this decrease in vital activity. Sometimes, however, its elevation can have the same consequences. We have already seen that germinating seeds, and among animals, frogs, become dormant at a high temperature; there is moreover an American mammal, the tanrec, which, it is said, goes into a true state of lethargy under the influence of high heat.

The highest vertebrates (warm-blooded animals), which have a more perfect *internal environment,* that is to say, circulating fluids in which the temperature is constant, are not subject to this influence by the external environment. Nonetheless, at a certain period of their existence, at the beginning, they start by being creatures with oscillating lives. This happens when they are in the state of the *egg.* The developmental work that must take place in the bird's egg requires a certain level of temperature, close enough to that of the adult animal; if this appropriate temperature is not provided for the egg, it remains dormant. It is not in a state of chemical indifference, because it can be established that it respires; it absorbs oxygen and gives off carbonic acid. Nevertheless this material exchange it not very active. A new-laid hen's egg is taken and placed in a footed cylinder above a layer of baryta water; this latter slowly becomes cloudy from the deposit of barium carbonate resulting from the exhalation of respiratory carbonic acid. The egg is able to remain for a certain time in this state of dormant life, ready to develop into a new animal if the conditions for incubation are realized. But it is not able to retain this aptitude indefinitely; after a few weeks it will become what we call *passé,* that is to say dead and unfit for incubation. It was therefore not completely inert; it lived obscurely.

When on the contrary the egg is subjected to a temperature of 38 to 40 degrees, vital activity will be increased; respiration, evidence of this energetic activity, will become very marked; the cicatricula will divide and proliferate; the rudiments of the embryo will appear at first, and by means of a successive epigenesis will complete the pattern of a fully formed bird; then life is no longer dormant; it is on the contrary extremely active.

We ought to ask ourselves how dormancy is produced under the influence of cold, and by what mechanism the return of heat imparts a new impulse to the vital activity. Experimentation establishes that the animal falls into a state of dormancy or hibernation because all its organic elements are surrounded by a cooler environment in which chemical actions, and proportionally, the functional manifestations of life are diminished. In the cold-blooded or hibernating animal mechanisms are lacking to maintain a constant environment around the elements, despite atmospheric variations. It is the cooling of the internal environment that benumbs the animal; it is the rewarming of this same environment that arouses it.

When a cold-blooded animal, a frog for example, has become dormant, it might be believed that the action of the cold is exerted primarily on its sensibility, on its nervous system, which is the general regulator of the functions of organic life and of animal life.

This is not so at all. When the *internal environment,* that is to say, the whole of the circulating fluids, is cooled, every element in contact with the blood becomes dormant on its own account, revealing thus its autonomy and the conditions of its own activity.

In a word, every organic system, every element, is in itself influenced by cold like the whole organism. It has the same conditions of activity or inactivity as the whole and it forms a new microcosm within the living being, itself a microcosm within the universe.

In the same way, when the dormant organism comes back to life it is not the nervous system that revives the other systems; how could this be since it is in the same state of dormancy as they are? It is again the *internal environment* that receives the influence of the *external environment,* and reawakens each element in turn, according to its sensibility or excitability. An experiment that I have carried out before places these ideas in clear relief. A frog is taken, dormant from the cold. Sensibility and motion are absent; the apparatus of organic life functions obscurely; blood returns red from the tissues where vital combustion is extremely attenuated; the heart gives but four beats a minute instead of fifteen or twenty as it does during the summer.

This frog can be brought out of its lethargic state. For this it is enough to rewarm it. How then does elevation of the temperature operate? It is not at all, we have said, by a nervous action exerted upon sensibility. To assure myself of this I carried out the following experiment: The paw of a dormant frog is placed in warm water, after its heart has been exposed. Whether the nerve to this limb has been sectioned, or whether it remains intact, the frog revives within the same time. The heart resumes its more rapid beats, and all the systems reawaken in succession. It is the rewarmed blood which has created around all the elements the physical condition of temperature necessary for vital activity. The blood, returning warmer from the paw, has revived the heart beats, and it is the excited heart which has aroused the animal.

The influence of temperature is thus clearly brought to light. In the frog one sees an animal with an oscillating life, that is to say, dependent upon the cosmic environment. A fall in temperature diminishes its vital activity; an elevation of temperature increases it.

The proposition, expressed in these terms, is too absolute, however. In this regard we ought to recall the facts that I have already mentioned to demonstrate that there are measures, gradations, and infinite nuances in the actions of physicochemical agents upon the organism. In a general way, it is true that by increasing the temperature vital activity is enhanced, but if the temperature exceeds certain limits, if for example in the frog it reaches 37 to 40 degrees, the animal is on the contrary anesthetized and stupefied. It is the same for seeds, which, excited to germinate at 20 degrees, are inactivated at 35 degrees. We place before you two frogs, the one which we have immersed in water at 37 degrees; you see that it is stupefied and no longer moves; it is in the same state as the second that had been placed in ice water. Let us change their jars; they will both revive except that it is the cold that revives the first, and the heat that revives the second.

Dormant or anesthetized animals and plants resist agents which would kill them if they were in a more active state of life. This resistance varies, moreover, with the nature of the toxic agents employed.

Because of the depression of their vitality, dormant animals resist conditions in which others would perish. Thus dormancy is also a condition of vital resistance as is latent life. A frog goes the entire winter without taking nourishment; the attenuation of the vital process permits this long suspension of the supply of materials, the animal could not support abstinence for such a long time if it were at a higher temperature. A very small bird, whose vital activity is always considerable, dies of hunger if it is left without food for twenty-four hours.

In their excellent studies on respiration, Regnault and Reiset have called attention to the remarkable resistance of hibernating marmots to conditions that would cause them to perish if they were in their ordinary state of life. A marmot, which breathes weakly during hibernation, can be placed without any inconvenience in an atmosphere deficient in oxygen; awakened, it would soon die from asphyxia. Similarly, this animal, which had remained for several months without food, and withstood the deprivation without harm, would no longer be able to endure it once it is awakened. It must be provided with abundant food, which it will devour with voracity, and without which it would soon die. I have often repeated this experiment, in dormice or marmots which I reawakened; if I did not give them food they soon succumbed, having rapidly exhausted the reserves derived from previous feeding.

To complete the account of the facts relating to *oscillating* life we can say that the mechanism of dormancy and the mechanism of the return to active life are as clearly explained as in the case of latent life.

The influence of cosmic conditions at first produces the incomplete suppression of the physical and chemical phenomena of vital destruction. Dormant animals no longer make any movements, their muscles undergo only a slight combustion, and their venous blood is nearly as bright as arterial blood; moreover, combustions are considerably reduced in the other tissues, heat production is weak, and carbonic acid is given off in small quantities. Thus it is the functional manifestation of life, which corresponds to the destruction of organs, that is primarily attenuated. Creative life undergoes a parallel reduction. It can even be said that it is entirely suspended as regards the formation of the immediate principles that constitute the reserves. Nonetheless, certain morphological phenomena, cicatrization and redintegration, still proceed very actively. We shall have to explain these facts at a later time.

The return to vital activity is further explained in the same way as reviviscence.

It is necessary for the hibernating animal to have reserves, not only to provide for the early expenditures of awakening, but also to suffice for the consumption it makes during the state of dormancy. Vital destruction, in fact, is not suspended, it is only diminished; as for vital creation, and the formation of reserves, it no longer has materials upon which to operate during hibernation, since the animal no longer takes in food from the outside.

This is why, before falling into the winter sleep, or as soon as they have a presentiment of its approach, animals prepare reserves in diverse forms. In the marmot the tissues load themselves with fat and glycogen; in the frog, and in all animals, organic provisions of various substances are accumulated. It is upon these prudent reserves, prepared by nature, that the animal lives during the period of dormancy; it no longer does anything but consume, it no longer creates, it no longer accumulates. These reserves suffice for a certain time for the attenuated manifestations observed in these dormant animals, but they would be exhausted rapidly if the vital activity were renewed. Thus it is necessary that as soon as they awaken the animals find at their disposal the foodstuffs on which the creative elaboration will be exerted. Dormice place within the nest in which they hibernate provisions that they consume as soon as they awaken. I have had the opportunity of making some interesting experiments on these animals. If dormant dormice are taken and sacrificed while they are fully asleep, and their liver is analyzed, a certain store of glycogen is still found therein, but if they are not sacrificed until four or five hours after they are awakened, hardly any traces of this material are found. These four hours of active life have used up the savings that would still have sufficed for several weeks of dormant life.

Besides the prolonged dormancy about which we have just spoken, and which the animal supports only on the condition of having considerable reserves accumulated beforehand, there are more or less passing dormancies that really do not require such reserves. Insects that are dormant in the morning after a cool night can be in full activity in the daytime sun. The immobile bee that can be picked up with impunity in the morning, is in a state in which it will sting vigorously toward noon. It is clear that these periods of activity and dormancy are too short, and succeed each other too rapidly, to require any considerable reserves; nevertheless, one can be satisfied that the great law of nutrition by means of reserved is immutable and that things take place more or less in the same way in all the states of life.

III. CONSTANT OR FREE LIFE

Constant, or free, life is the third form of life; it belongs to the most highly organized animals. In it, life is not suspended in any circumstance, it unrolls along a constant course, apparently indifferent to the variations in the cosmic environment, or to the changes in the material conditions that surround the animal. Organs, apparatus, and tissues function in an apparently uniform manner, without their activity undergoing those considerable variations exhibited by animals with an oscillating life. This is because in reality the *internal environment* that envelops the organs, the tissues, and the elements of the tissues does not change; the variations in the atmosphere stop there, so that it is true to say that *physical conditions of the environment* are constant in the higher animals; it is enveloped in an invariable medium, which acts as an atmosphere of its own in the constantly changing cosmic environment. It is an organism that has placed itself in a hothouse. Thus the perpetual changes in the cosmic environment do not touch it; it is not chained to them, it is free and independent.

I believe I was the first to insist upon this idea that there are really two environments for the animal, an *external environment* in which the organism is placed, and an *internal environment* in which the elements of the tissues live. Life does not run its course within the external environment, atmospheric air for the air breathing creatures, fresh or salt water for the aquatic animals, but within the *fluid internal environment* formed by the circulating organic liquid that surrounds and bathes all of the anatomical elements of the tissues; this is the lymph or plasma, the liquid portion of the blood which in the higher animals perfuses the tissues and constitutes the ensemble of all the interstitial fluids, is an expression of all the local nutritions, and is the source and confluence of all the elementary exchanges. A complex organism must be considered as an association of *simple beings,* which are the anatomical elements, and which live in the fluid internal environment.

The constancy of the internal environment is the condition for free and independent life: the mechanism that makes it possible is that which assured the maintenance, within the *internal environment,* of all the conditions necessary for the life of the elements. This enables us to understand that there could be no free and independent life for the simple beings whose constituent elements are in direct contact with the cosmic environment, but that this form of life is on the contrary the exclusive attribute of beings that have arrived at the summit of complication or organic differentiation.

The constancy of the environment presupposes a perfection of the organism such that external variations are at every instant compensated and brought into balance. In consequence, far from being indifferent to the external world, the higher animal is on the contrary in a close and wise relation with it, so that its equilibrium results from a continuous and delicate compensation established as if by the most sensitive of balances.

The conditions necessary for the life of the elements which must be brought together and maintained constant in the internal environment for the exercise of free life, are those that we know already: water, oxygen, heat, and chemical substances or reserves.

These are the same conditions as those which are necessary for life in the simple beings, except that in the more perfect animals with independent life, the nervous system is called upon to regulate the harmony among all these conditions.

1. Water

This is an indispensable element, qualitatively and quantitatively, in the constitution of the environment in which the living elements function and evolve. In free-living animals there must exist an ensemble of dispositions regulating output and intake so as to maintain the necessary quantity of water within the internal environment. In the lower beings the quantitative variations in water compatible with life are more extensive, but the creature is on the other hand without means of regulating them. This is why it is chained to the vicissitudes of the climate, dormant in latent life during dry weather, revived in wet weather.

The highest organism is inaccessible to hygrometric variations, thanks to artifices of construction and to physiological functions which tend to maintain the relative constancy of the quantity of water.

In man, especially, and in general in the higher animals, loss of water occurs in all the secretions, in the urine and the sweat especially, and secondarily in respiration, which carries off a notable quantity of water vapor, and finally by cutaneous perspiration.

As to the intake, this is accomplished by the ingestion of fluids or of foods that include water, or even in some animals by absorption through the skin. In all events, it is most likely that the whole quantity of the water in the organism comes from the outside by the one or the other two routes. It has not been possible to demonstrate that the animal organism really produces water; the contrary opinion appears to be nearly certain.

It is the nervous system, we have said, that provides the mechanism for compensation between intake and output. The sensation of thirst, which is under the control of this system, makes itself felt whenever the proportion of fluid diminishes within the body as the result of some condition such as hemorrhage or abundant sweating; the animal thus finds itself induced to drink in order to restore the losses it has undergone. But even this ingestion is regulated in the sense that it cannot increase the quantity of water present in the blood beyond a certain level; urinary and other excretions eliminate the surplus as a sort of overflow. The mechanisms that vary the quantity of water and reestablish it are thus most numerous; they set in motion a host of mechanisms of secretion, exhalation, ingestion, and circulation which transport the ingested and absorbed fluid. These mechanisms are varied, but cooperate to the same end: thc presence of water in effectively fixed proportions within the internal environment, the condition for free life.

These compensatory mechanisms exist not only for water; they are observed also for most of the mineral and organic substances contained in solution in the blood. It is known that the blood cannot take on a considerable load of sodium chloride, for example; above a certain limit the excess is eliminated in the urine. As I have established, it is the same for sugar, which normally present in the blood, is, above a certain quantity, eliminated in the urine.

2. *Heat*

We know that for each organism, elementary or complex, limits of external temperature exist between which its activity is possible, with a midpoint which corresponds to the maximum of vital energy. This is true not only for beings that have arrived at the adult state but also for the egg or embryo. All these creatures are subject to oscillating life, but for the higher animals, the so-called warm-blooded animals, the temperature compatible with the manifestations of life is closely fixed. This fixed temperature is maintained within the internal environment despite extreme climatic variations, and assures the continuity and the independence of life. In a word, there exists in animals with constant and free life a function of calorification which does not exist at all in animals with an oscillating life.

For this function there exists an ensemble of mechanisms governed by the nervous system. There are *thermic* nerves and *vasomotor* nerves to which I have called attention, whose activity produces sometimes an elevation and sometimes a fall in temperature, according to the circumstances.

The production of heat is due, in the living world as in the inorganic world, to chemical phenomena; such is the great law whose understanding we owe to Lavoisier and Laplace. It is in the chemical activity of the tissues that the higher organism finds the source of the heat it conserves within its internal environment, at a nearly constant level, from 38 to 40 degrees for mammals, and 45 to 47 degrees for birds. The regulation of heat takes place, as

I have said, by means of two kinds of nerves; the nerves that I call *thermic,* which belong to the sympathetic system and serve as brakes, so to speak, on the chemicothermic activities taking place within the living tissues. When these nerves act, they diminish the interstitial combustions and lower the temperature; when their influence is weakened by suppression of their action or by the antagonism of other nervous influences, then combustions are increased, and the temperature of the internal environment rises considerably. The *vasomotor* nerves, by accelerating circulation in the periphery of the body, or in the central organs, intervene also in the mechanisms for the equilibration of animal heat.

I will add only this last fact. When the action of the cerebrospinal system is considerably attenuated, while that of the sympathetic *(thermic nerve)* is permitted to remain intact, temperature is seen to fall considerably, and the warm-blooded animal is so to speak converted into a cold-blooded animal. I have carried out this experiment on rabbits, cutting the spinal cord between the seventh cervical vertebra and the first dorsal. When, on the contrary, the sympathetic is destroyed, leaving the cerebrospinal system intact, the temperature is noted to rise, at first locally and then generally; this is the experiment I carried out in horses by cutting the sympathetic trunk, especially when they were weakened beforehand. A true fever then follows. I have elsewhere developed the history of all these mechanisms at length (see Leçons sur la Chaleur Animale, 1873); I recall them here only to establish that the calorific function characteristic of warm-blooded animals results from the perfecting of the nervous mechanism which, by an incessant compensation, maintains an apparently fixed temperature within the *internal environment,* within which there live the organic elements to which ultimately we must always attribute all the vital manifestations.

3. Oxygen

The manifestations of life require for their production the intervention of air, or better, its active portion, oxygen, in a dissolved form and in an appropriate state for it to reach the elementary organism. It is moreover necessary for this oxygen to be in proportions that are to a certain degree constant within the internal environment; too small a quantity or too great a quantity are equally incompatible with the vital functions.

Thus in animals with a constant life appropriate mechanisms are required to regulate the quantity of this gas which is assigned to the internal environment, and to keep it more or less constant. In the highly organized animals the penetration of oxygen into the blood is dependent upon the respiratory movements and the quantity of this gas present in the ambient environment. Moreover, the quantity of oxygen that is found in the air depends, as physics teaches it, on the percentage composition of the atmosphere and its pressure. Thus it can be understood that an animal could live in an atmosphere less rich in oxygen if an increase in pressure compensated for this decrease, and inversely, that the same animal could live in an environment richer in oxygen than ordinary air if a diminution in pressure compensated for the increase. This is an important general proposition, resulting from the work of Paul Bert. It can be seen in this case that the variations in the environment compensate and balance each other, without the intervention of the animal. If the percentage composition diminishes or increases in the opposite direction, when the pressure rises or falls, the animal ultimately finds the same quantity of oxygen in the environment and its life goes on under the same conditions.

But there can be mechanisms within the animal itself that establish this compensation when it is not accomplished on the outside, and which insure the penetration into the internal environment of the quantity of oxygen required by the vital functions; we would mention the different variations that can take place in the quantity of hemoglobin, the active absorbing material for oxygen, variations that are still little known but which certainly also intervene for their own part.

All these mechanisms, like the preceding, are without effect except within rather restricted limits; they are perverted and become powerless in extreme conditions. They are regulated by the nervous system. When the air becomes rarefied for some reason, such as during ascension in a balloon or on mountains, the respiratory movements become deeper and more frequent, and compensation is established. Nevertheless, mammals and man cannot sustain this struggle for compensation very long when rarification is extreme, as when for instance they are transported to altitudes above 5000 meters.

We cannot enter here into the particular details that the question deserves. It suffices for us to propose it. We call attention only to an example related by Campana. It is relative to the high-flying birds, such as the birds of prey and particularly the condor, which rises to heights of 7000 to 8000 meters. They remain there, moving around for long periods of time, although in an atmosphere that would be fatal to a mammal. The principles set forth above permit the prediction that the internal respiratory environment of these animals ought to escape, by some appropriate mechanism, from the depression of the external environment; in other terms, that the oxygen contained in their arterial blood ought not to vary at these great heights. In fact there are in the birds of prey enormous pneumatic sacs, connected to the wings, which do not operate except when these move. When the wings lift, they are filled with external air, when they fall, they pump the air into the pulmonary parenchyma. So that, as the air is rarified, the work of the bird's wing which supports it is necessarily increased and consequently the supplementary volume of air passing through the lungs is also increased. The compensation for the rarification of the external air by an increase in the quantity inspired is thereby assured, and with it the constancy of the respiratory environment characteristic of the bird.

These examples, which we could multiply, demonstrate to us that all the vital mechanisms, however varied they might be, always have one purpose, that of maintaining the integrity of the conditions for life within the internal environment.

4. Reserves

Finally, it is necessary for the maintenance of life that the animal have reserves that assure the constancy of the constitution of its internal environment. Highly organized beings draw the materials for their internal environment from their food, but as they cannot be subjected to an identical and exclusive kind of diet, they must have within themselves mechanisms that derive similar materials from these varied diets and regulate the proportion of them that must enter the blood.

I have demonstrated, and we shall see later, that nutrition is not *direct* according to the teaching of accepted chemical theories, but that on the contrary, it is *indirect* and carried out by means of reserves. This fundamental law is a consequence of the variety of the diet as compared with the constancy of the environment. In a word, *one does not live by his present food, but by that which he has eaten previously,* modified, and in some way created by assimilation. It is the same with respiratory combustion; nowhere is it *direct,* as we shall demonstrate later.

Thus there are reserves, prepared from the food, and consumed at each moment in greater or lesser proportions. The vital manifestations thus destroy the provisions which no doubt have their primary origin from the outside, but which have been elaborated within the tissues of the organism, and which, added to the blood, insure the constancy of its chemicophysical constitution.

When the mechanisms of nutrition are disturbed, and when the animal finds it impossible to prepare these reserves, when it only consumes those that it had accumulated beforehand, it is on its way to ruin, that can end only in the impossibility of life, in death. It would then be of no use for it to eat; it would not be nourished, it would not assimilate, it would waste away.

Something of the kind takes place when the animal is in a state of fever; it uses without restoring, and this state becomes fatal if it persists to the complete exhaustion of the materials accumulated through previous nutrition.

Thus, the nutritive substances that enter an organism, whether animal or plant, do not participate in nutrition directly or immediately. The nutritive phenomenon takes place in two stages and these two stages are always separated from one another by a longer or shorter period, whose duration is a function of a host of circumstances. Nutrition is preceded by a particular elaboration that is terminated by a *storage of reserves* in the animal as well as in the plant. This fact permits one to understand how a being can continue to live, sometimes for a long time, without taking food; it lives on its reserves, accumulated within its own substance; it consumes itself.

These reserves are of variable importance depending upon the creatures concerned, and the various substances, in different animals and plants, and in annual or biennial plants, etc. This is not the place to analyze such a vast subject; we have wanted to show that the formation of reserves is not only the general law of all forms of life, but that it constitutes also an active and indispensable mechanism for the maintenance of a constant and free life, independent of variations in the ambient cosmic environment.

Conclusion

We have examined in succession the three general forms in which life appears: *latent* life, *oscillating* life, and *constant* life, in order to see whether in any of them we might find an internal vital principle capable of producing its manifestations, independently of external physicochemical conditions. The conclusion to which we find ourselves led is easy to draw. We see that in latent life the being is dominated by external physicochemical conditions, to the point that all vital manifestations can be arrested. In oscillating life, if the living being is not as absolutely subject to these conditions, it nevertheless remains so chained to them that it is subject to all their variations. In constant life the living being seems to be free, and vital manifestations appear to be produced and directed by an inner vital principle free from external physicochemical conditions; this appearance is an illusion. On the contrary, it is particularly in the mechanism of constant or free life that these close relations exhibit themselves in their full clarity.

We cannot therefore admit the presence of a free vital principle within living beings, in conflict with physical conditions. It is the opposite fact that is demonstrated, and thus all the contrary concepts of the vitalists are overthrown.

THIRD LECTURE

Classification of the Phenomena of Life

SUMMARY: I. Classification of the phenomena of life—Two great groups: organic destruction and creation—This classification characterizes general physiology and embraces within its confines all the manifestations of life—Unity of life in the two kingdoms.

II. Classifications of the Living Beings; Linneus, Lamarck, de Blainville—Theories of the duality of life in two kingdoms—Differentiation of the natural kingdom—Contrast between animals and plants—Chemical, physical, and mechanical antithesis between animals and plants—Priestley, Saussure, Dumas, and Boussingault, Huxley, Tyndall.

III. General Refutation of Dualistic Theories of Life in Animals and Plants—Latest form of the theory of the duality of life—The quality of life and general physiology—Unity of the laws of life; variety of the manifestations of life, and the different functioning of living machines—Conclusion: the integrity of the phenomena of organic destruction and creation proves the unity of life.

I. We have demontsrated in living beings two characteristic aspects of their existence; *life,* organic creation, and *death,* organic destruction. Today it is a question of affirming this division and of showing that it serves as the basis for general physiology. We shall consider here the characteristics of life only in their essence and in their universality, and from this point of view we shall classify them in two major divisions:

1. Phenomena of wear and tear, of *vital destruction,* which correspond to the functional phenomena of the organism.

2. *Plastic* phenomena, or those of vital creation, which correspond to functional rest and organic regeneration.

Everything that takes place in the living being relates to either one or the other of these types, and *life is characterized* by the union and interlinking of these two orders of phenomena. This division of the phenomena of life seems to us to be the best of those that could be proposed in general physiology. It is altogether the most inclusive and the most consistent with the real nature of things. Whatever forms life might assume, and whatever the complexity or the simplicity of these forms, the preceding classification is applicable to them. We could not conceive of any living being, even any living particle without the interplay of these two orders of phenomena. This is the physiological foundation around which all the varieties of life in the two kingdoms revolve.

The classifications of the phenomena of life that have been proposed up to now apply to the higher animals and relate especially to descriptive physiology; they are far from expressing this generality.

In general physiology a classification must correspond to the phenomena of life, independent of the morphological complexity of the beings, and should be based solely upon the universal properties of living matter, abstracted from the specific moulds into which it has entered. It is precisely this condition that is satisfied by the division into phenomena of *organic destruction* and *creation.*

Before studying in the remainder of the course each of these phases of vital activity, organic *destruction* and organic *creation,* it is important to clarify and firmly establish, beginning with this lecture, the close relationship that indissolubly unites the two aspects of our classification of the vital phenomena. This classification is the expression of life at once in its broadest and in its most detailed form. It is applicable to all living beings without exception, from the most complicated organism of all, man, to the simplest elementary being, the living cell. In a word, one cannot conceive otherwise of a being endowed with life.

In fact, these phenomena take place simultaneously in every living being, in an enchainment that cannot be broken. Disorganization or dissimilation makes use of the living material in the *functioning* organs; the assimilative synthesis regenerates the tissues, it collects the reserve materials that activity is to consume. These two operations, of destruction and reconstruction, the one the opposite of the other, are absolutely connected and inseparable, in this sense, at least, that destruction is the necessary condition for reconstruction. The phenomena of functional destruction are themselves the precursors and the initiators of the material reconstruction by the formative process that operates silently within the intimacy of the tissues. Losses are restored as they take place and equilibrium is reestablished as soon as it tends to be disturbed; the body maintains its own composition. This wear and tear, and this renovation of the constituent parts of the organism, as we saw at the beginning of this course, make existence nothing more than a perpetual alternation of *life* and *death,* of composition and decomposition. There is no life without death; there is no death without life.

Moreover, such a classification has absolutely nothing unusual about it; it does not constitute, properly speaking, a novelty in science. Everyone has more or less perceived these two aspects of the vital activities, and we have cited as examples numerous passages in the essays on the definition of life that we recalled in our first lecture. The essential point is to have understood the importance and all the implications of this simple and fruitful classification, and to bring out all its consequences.

Eighty years ago, Lavoisier had clearly perceived the two phases of vital activity; the *disorganization* or destruction of animal or plant organism by combustion and putrefaction, and the organic creation, *vegetation* or *animalization,* which are operations opposite to the former.* "Since," he said, "combustion and putrefaction are the means nature employs to return to the mineral kingdom the materials she had taken from it to form the plants and animals, vegetation and animalization must be operations opposite to combustion and putrefaction."

***Pièces historiques concernant Lavoisier,* communicated by Dumas *(Leçons de Chimie Professées à la Société Chimique de Paris).* Paris, 1861, p. 295.

It is thus not possible to separate in any living being these two modalities of life that are encountered in plants as well as in animals.

Here, then, is a physiological axiom that implies the unity of life; we shall formulate it at the beginning; we shall see it verified in the entire course of our studies, and it will serve as a criterion for judging the various theories in which the life of plants is placed in opposition to that of animals.

In fact, contrary to the principle that we have just enunciated, and which constitutes, let us repeat, the *axiom* of general physiology, several famous theories have stated that the two orders of vital phenomena, instead of pertaining to all living beings, were to be attributed to different beings, some belonging to the animal kingdom, the others to the plant kingdom.

These theories of the division of the two vital factors between the two kingdoms, which can be called theories of *vital duality,* are contradicted by our principle, and we can add, by examination of the facts. There is no single category of being responsible for *organic synthesis* and another category for combustion, or *organic analysis.* As we have said, there cannot be life except where there is both organic synthesis and organic destruction.

General physiology must examine these viewpoints at their origins and in the different forms they have assumed. In France it was Dumas and Boussingault, in Germany Liebig, and in England Huxley* and Tyndall, who created and propagated these diverse theories in science. In recalling them, we ought to give credit to the simplicity and the breadth of the views by which their authors supported them, and recognize the service that they rendered in stimulating a considerable number of investigations, works, and discoveries. We shall see moreover that our difference of opinion comes from a difference in point of view. The creators of the dualistic theories have considered the two factors of life, in their relation to the cosmic environment, without being as impressed as we by the identity of their origin and their indissoluble unity.

It has been thought possible to attribute to Lavoisier the first idea of this duality; but the writings of this illustrious founder of

*Huxley, *La Place de l'Homme dans la Nature. Paris, 1868, and Les Sciences Naturelles et les Problèmes qu'elles font Surgir.* Paris, 1877.

modern chemistry that have been invoked do not seem to me to be conclusive in this sense. We have already quoted above a passage in which Lavoisier recognizes the existence in the living beings of these two inverse phenomena by which they carry out the *synthesis* of the organism (animalization, vegetation,) and on the other hand its *destruction* (combustion, fermentation, putrefaction).

Lavoisier does not at all separate animals from plants in this regard; he seems to consider that they behave in a similar manner with respect to the mineral kingdom, and nowhere does he say that the vegetable kingdom must serve as the exclusive intermediary between the mineral kingdom and the animal kingdom.

Thus it is not to Lavoisier that the theory of the chemical antagonism between animals and plants can be attributed; it seems to us that the germ of it exists in older works, and in particular in the celebrated studies of Priestley on the antagonism in the respiration of animals and plants.

Moreover, it must indeed be said, this idea of an opposition between the two kingdoms must have existed at all times, because it results from the appearance of things, and appearances have always deceived us as to the real nature of phenomena. There is in fact a morphological distinction between animals and plants that is clearly enough marked externally for one to have thought that it was profoundly inscribed in the organization and manifestations of life. But this distinction is only in form, on the surface, and not at the base of the phenomena. As for ourselves, we maintain that there is an identity in the essential attributes of life in the two kingdoms, and that the division we have established in the acts of life, *destruction* and vital *creation,* apply universally to living beings. To justify this fundamental division which we have introduced into general physiology it is necessary first to set forth the contrary theories and refute them in their principal points.

II. CLASSIFICATIONS OF LIVING BEINGS AND DUALIST THEORIES OF LIFE

Natural beings were first divided into two great empires; the one, formed by animate beings, and the other by inanimate beings. This distinction was made by Aristotle. It was only later, about 1645, that a French alchemist named Colleson first formulated the division of nature into three kingdoms, animal, vegetable, and mineral, which included all terrestrial objects; for sidereal bodies he invoked a fourth kingdom, the planetary kingdom. In each of these domains there existed an ideally perfect type, a king; man among the animals, the grape among the plants, gold for the minerals, and the sun for the celestial bodies.

The division into these kingdoms having thus been born, Linneus* sanctioned it by giving it the three following characters:

Esse	*Vivere*	*Sentire*
mineral	plant	animal

He expressed them also in the following formula:

Mineralia sunt
Vegetalia sunt et crescunt
Animalia sunt, crescunt et sentiunt

There are naturalists, de Blainville for example, who, placing man above all of the animals, have formed a special kingdom for him, the *kingdom of man,* characterized by one further attribute, *intelligence; homo intelligit.*

Lamarck, however, resumed the binary division, and at first not distinguishing at all between living beings, recognized two classes of bodies:

Living bodies
Brute or inanimate bodies

Nevertheless the division into three realms prevailed, and the two kingdoms of animals and plants were considered almost as separate from one another as each one of them was from the mineral kingdom. To create distinct categories for animals and plants, we certainly do not oppose in any way, but to go on from there to establish between the two groups of beings a difference so profound that it would involve, in a way, two different physiologies, the one animal, the other plant, based on special principles, becomes a point of view that we must oppose.

The elements of differentiation between the modalities of life in animals and plants were first sought in anatomy. Cuvier, to

*Linneus, *Systema Naturae.* Editio prima, reedita cura A.L.A. Fée. Parisiis, 1830.

cite but one example, called attention to the lack of a digestive apparatus in plants as a very general characteristic that could serve to distinguish them from animals. Today it is well known that a large number of lower animals have no digestive tube, and that at higher levels the males of certain species such as the Rotifers have been deprived of them while the females possess them. In fact, this characteristic has therefore no absolute value at all; in principle, as we shall see later on, the digestive apparatus is only an accessory apparatus in nutrition. The reserves, which are in reality the nutritive resource of living beings, are identical in animals and in plants.

In the second place a difference has been thought to be found between animals and plants from the point of view of the composition of their tissues.

It has been said for example that nitrogen is an element characteristic of the animal organism, while it exists only exceptionally in plants. Analysis of the parenchyma of mushrooms and of the seeds of phanerogamous plants quickly upset this opinion. It is accepted today that protoplasm, the sole active and functional portion of the plant, has the same composition as animal protoplasm; it is a nitrogenous substance. Nitrogen, instead of being an accessory element, is therefore essential and fundamental in the two kingdoms. The anatomic elements of plants, the cells, fibers, and vessels in certain regions lose their protoplasm and no longer intervene in the constitution of the plant except as supporting parts. To a lesser degree this is encountered in animals; the skeleton of crustaceans and the carapace of insects are parts that are poor in nitrogen or are even absolutely lacking in it. The principal substance of the supporting tissues in plants is *lignin* or *cellulose.* But the proposition has been advanced that cellulose was special to plants and belonged to them alone. This is not so. This substance has been encountered in the envelope of tunicates, and close analogies have in addition been established with the *chitin* that forms the carapace of crustacea and insects.*

Nevertheless, as we have said, it is in the relations of animals

*C. Schmidt. *Zur Vergleichenden Physiologie der Wirbellosen Thiere,* 1845.—Berthelot, *Comptes rendus de la Sociéte de Biologie.*

and plants with the atmosphere that the theory of Dualism has found its first and its strongest arguments. The discoveries made on this subject at the end of the last century at once placed the life of plants in opposition to that of animals.

The celebrated experiment of Priestley is well-known, by which this great chemist established that plants purify the air that animals have vitiated, and seem to behave, in regard to their respiration, in an opposite sense. A mouse is placed under a bell-jar in confined air; it finally dies, the air is vitiated, and if another animal is introduced, it very rapidly succumbs and in its own turn dies of asphyxia. But, if a plant (a sprig of mint) is placed in the bell-jar, the atmosphere is purified, its initial constitution is reestablished, and an animal can again live within it.†

The plant being thus lives where the animal dies; they behave in precisely the opposite manner in relation to the environment, the one undoing what the other has done, and between the two of them they constitute a harmonious, balanced, and consequently enduring state of affairs.

This experiment was really the starting point for the modern chemical opposition of animals and plants. Animals absorb oxygen and exhale carbonic acid. Subsequent investigations by Ingen-Housz, Sénébier, and Th. de Saussure have proven that in the green parts of plants, under the influence of the sun's rays, there takes place on the contrary an absorption of carbonic acid and an exhalation of oxygen.

This contrast between the respiration of animals and that of plants has been generalized in an imposing way by Dumas and Boussingault in their theory of *cycling of materials* between the two organic kingdoms:

> The oxygen removed by the animals is restored by the plants. The former consume oxygen, the latter produce oxygen. The former burn carbon, the latter produce carbon. The former exhale carbonic acid, the latter fix carbonic acid.

The animal was thus considered to be an apparatus for *combustion,* for *oxidation,* for *analysis,* or for *destruction,* while on

†See Priestley, *Experiences sur les Airs,* vol. III.

the contrary, the plant was an apparatus for *reduction,* for *formation,* for *synthesis.*

It resulted from this that the phenomena of destruction or vital combustion were absolutely separated in living beings from the phenomena of reduction or organic synthesis. Vital creation was vested in plants while organic destruction was reserved for animals. The animal organism, being incapable of forming any of the principles that enter into its constitution, such as fat, albumin, fibrin, starch, and sugar, all were provided for it by the plant kingdom, and the nutrition of animals was nothing but the putting in place of materials elaborated solely by the plants. The milk secreted by the herbivors, as well as the casein, the butter, and the sugar ought to be recoverable, pound for pound, from the plants upon which they fed, etc.

These ideas were again collected and expressed with a luminous simplicity by Dumas and Boussingault, in their chemical balance of living beings. We reproduce there the striking formula of this celebrated theory:

A plant:	*An animal:*
Produces sugar, fatty and albuminoid materials	*Consumes* sugar, fatty and albuminoid materials
Reduces, with liberation of oxygen:	*Produces,* with absorption of oxygen:
CO_2	CO_2
HO	HO
NH_4O	NH_4O
Absorbs heat	*Gives off* heat
Is *immobile*	*Moves*

This is to say in other terms that *formation or chemical synthesis* belongs to plants and that *combustion* belongs to animals.

Now this conclusion is contradictory to the fundamental principle of general physiology, namely, that the two phases of vital action, *creation* and *destruction,* instead of being divided between the two realms, are intimately united in every being and in every living part.

But the duality of life is not affirmed only from the chemical point of view; it has in our own times assumed another form that we can call *dynamic* or *mechanical.*

The human body and that of animals has often been compared with a combustion engine. The chemists have established that the products given off by the body, the excretions taken as a whole, contain a much greater proportion of oxygen than the ingested foods. Thus a continual combustion takes place in the animal organism, a source of *heat* and *mechanical energy.* Says Huxley:

> The oxidation of complex compounds, which enter the organism, is ultimately proportional to the sum of the energy that the body expends, exactly in the same way that the sum of the work that is obtained from a steam engine, and the quantity of heat that it produces are in strict proportion to the amount of coal that it consumes.
>
> The particles of matter that enter into the vital whirlpool are more complicated than those that emerge from it. To employ a metaphor that is not without some reality, the atoms that enter into the organism are for the most part fashioned in large masses, and break up into small masses before leaving it. The energy set free in this fragmentation is the source of the active forces of the organism.

From this comes the comparison of the animal body to a steam engine in which active energy is engendered. The organism, it has been said, is a machine, and is even quite perfect because, for a certain quantity of combustibles it produces twice as much work as the most economical engine. Its output, according to Moleschott, would amount to a fifth of the mechanical equivalent of the heat liberated by the combustion of the hydrogen and carbon that it consumes. Considering the two kingdoms from the point of view of the services that they render each other, as the adherents of final causes do, and not from the point of view of their essential activity, it has been said that the one was a reservoir of energy and the other a consumer. "The most complicated phenomena of life are summarized," Tyndall has said, "in this general law: The plant is produced by the lifting of a weight, the animal by the fall of this weight."

The plant would thus create energy like the mechanic who lifts the weight of a clock; by this action the work of the gear train is created in potential; it suffices to let the mass fall in order to manifest it. It is this that in mechanics is called potential energy, an energy of *tension*.

The plant would create tensile energies, and at the expense of the active energy of the sun. Under the influence of vibrations transmitted by the solar rays and by the warmth of the atmosphere, the chlorophyll (with which the plant kingdom is here equated) would separate the oxygen from its oxygenated combinations (water, carbonic acid, ammoniacal salts) that it absorbs. This oxygen, placed in the presence of combustible substances, is ready to combine with it, to create work in this way, and develop energy. The separation accomplished by the plant would be equal to the production of a potential energy, of tensile forces; the role of the plant kingdom would consist in transforming acting forces into forces of tension.

On the contrary, the animal would transform the *tensile* forces into *active forces*. The weight lifted by the plant, it lets fall; to return to our illustration, it releases the weight that moves the clock, it precipitates on the combustible substances the oxygen that the plant has separated.

What is necessary to accomplish this? According to Hermann, from whom we take this theory, it is necessary to destroy the obstacle that prevents the oxygen from combining, to remove the catch that holds the weight of the clock, to destroy, in short, the obstacle that prevents the tensile force from becoming an active force, or work; there must therefore be *releasing forces*.

Thus, *tensile forces* accumulated in the plants, *active forces* and forces of *release* in animals; this is the distribution that would constitute the dynamic duality of living beings.

III. GENERAL REFUTATION OF DUALIST THEORIES OF LIFE

To these theories general physiology can raise objections of principle and objections of fact. The great objection of principle that we address to the doctrine of vital duality is that it is in radical contradiction to our fundamental concept of life, which requires in all beings, animal or plant, the union of the phenomena of organic creation and destruction. We cannot conceive of a living being, animal or plant, apart from this formula; therefore we regard as erroneous *a priori* any proposition contradictory to this great physiological principle.

The second objection in principle that we shall formulate is relative to the idea of a *direct nutrition,* accepted by the dualist theory, and which physiology contradicts. The dualist theory supposes in fact that food passes directly from plants into animals, and that their immediate principles take their place therein, each according to its nature. Physiological study of the phenomena proves that nothing of the kind takes place, and that *nutrition is indirect.* The food first disappears, as a definite chemical material, and it is only after extensive organic work, after a complex vital elaboration, that the food comes to constitutes the *reserves,* always identical, that serve for the nutrition of the organism. Nutrition and digestion are completely separate; the nature of the food is essentially variable, and never has any effect in the normal state on the formation of the reserves, which remain constant, like the composition of the fluids and the organic tissues. In a word, the body never nourishes itself directly from the various foods, but always by means of identical reserves, prepared by a sort of work of secretion. And what we say here about the formation of *nutritive reserves* is found also in the two realms, in animals as well as in plants.

Moreover, it must be recognized, facts themselves have come to demonstrate that vital duality cannot exist in the absolute form that it has assumed.

As far as the formation of immediate principles is concerned, the question has been resolved, and the solution accepted even by those who had at first supported the contrary theory. It has been demonstrated that animals really form fats independently of those they ingest and can obtain from the food. The herbivor creates fat instead of finding it fully formed, and the carnivor acts in the same way. Not only do animals make fats, but they do not use directly that which is contained in their food. Nature does not recognize the kind of economy that would utilize ready-formed substances such as would come to our mind. It does not make use of a task already accomplished as if it were advantageous.

The dog, for instance, does not get fat on mutton fat, it makes dog fat. I have myself, with the collaboration of Berthelot, tried to produce an experimental demonstration of this fact by employing a means of recognizing and following the fat supplied to an animal; this method consists in using chlorinated fat as food, in which chlorine replaces some of the hydrogen. If the animal subjected to this diet presents a fat different from that which has been given to it, and has the characteristics appropriate to the organism that produced it, it would be necessary to conclude that no simple storage took place of the food that was introduced.

It could also be demonstrated that the albuminoid substances that constitute the animal tissues are not directly derived from the nutritive substances of plants.

But is is especially for the formation of sugars that doubts have been completely removed. About thirty years ago it was believed that sugar was, without any question, a plant substance, and that that which existed in animal organisms had of necessity been derived from plants. I succeeded in demonstrating that it is entirely otherwise, and that the animal itself manufactures this substance indispensable for vital functions, at the expense of a wide variety of food that it is given. I proved, moreover, that the sugar is produced in the animal by a mechanism identical to that which takes place in the plant.

We shall return to these facts with regard to the study of the phenomena of organic creation. Here let us conclude simply that with respect to the formation of immediate principles, experiment shows that animals and plants are not different, and that both the one and the other can form the same organic principles.

An antagonism between respiration of the animals and that of plants is not confirmed by either. The reduction of carbonic acid carried out by plants is part of the function of chlorophyll; this has no connection with respiration, which is identical in the two realms. Plant protoplasm, the uncolored parts, roots, seeds, etc. have the same respiratory properties as animal tissues. Like the animal, the plant absorbs oxygen, exhales carbonic acid, and produces heat; the fact is not to be doubted when the germination of seeds is observed.

Relative to the sensibility that would constitute the third point of antagonism between plants and animals, we shall have the opportunity of showing that it is in no way an exclusive attribute of animality (see Lecture VII). If plants do not exhibit locomotor functions comparable to those of animals, they nevertheless possess a sensibility, which is the *primum movens* of every vital act.

If the partisans of the chemicophysical opposition between animals and plants have had to yield to the evidence of contrary facts and renounce the arbitrariness of their old opinions, the spirit of the theory survives none the less; it is interesting to note that vital duality is concentrated now on a single argument.

It can no longer be doubted, we have said, that animals and plants are capable of producing the same immediate principles; it can no longer be denied that the one and the other are sites of destructions and reductions, infinitely numerous and interconnected. The difference between animals and plants would reside only in the agent or the energy that is the cause of the chemical or mechanical phenomena that take place within them. This is a point that we shall treat in greater detail, in the study of phenomena of vital creation (see Lecture VII). For the moment it will suffice to recall the major features of the question. It is admitted today* that the synthetic phenomena in plants and animals form two groups: those that require solar energy, which are reductions taking place in green plants under the influence of chlorophyll, and those which take place under the influence of combustion, occurring in animals or in those parts of plants that do not contain the green material. Such would be the two sources of active energy which accumulates in living beings; sometimes they are derived directly from solar energy, sometimes they are derived from the heat produced by combustion. The active energy comes from the sun when there is chlorophyll; in all other cases, whether in animals or in plants, it comes from the heat given off in oxidation or in chemical combinations of a similar kind. As an example of this latter type we can take brewer's yeast, *Saccharomyces cerevisiae*. This fungus contains no green matter at all; it has

*See Boussingault, *C.R.*, April 10, 1876, Vol. LXXXII, p. 788—*C.R.*, April 24, 1876.

no chlorophyll. Thus this plant cannot obtain its carbon directly from carbonic acid; it needs an *explosive* combustible material, sugar, that is to say one that can give off heat in burning. Here, thermal energy would replace solar energy.

The whole difference between living beings would finally be reduced to this.

We shall note that this new character cannot serve to distinguish animals from plants. Although plants are provided with chlorophyll, especially during the summer, in a manner incomparably more abundant than animals, it is not possible in any absolute way to equate the plant with chlorophyll. It should simply be said that there are beings containing chlorophyll and capable of using the energy emanating from the sun; this would be the realm of beings with chlorophyll; then would come the realm of beings without chlorophyll which are obliged to depend *indirectly* upon the sun for the dynamic force that they must use, that is to say on compounds definitively formed under the influence of its rays. But this classification, which would consist in ranking the beings according to the presence or absence of the green matter, chlorophyll, no longer corresponds to the classification of living beings into plants and animals. All of the vast class of the fungi, devoid of chlorophyll, would have to be removed from the plants, and many animals, (*Euglena viridis, Stentor polymorphus,* etc. etc.) would have to be included among the plants.

From the philosophical point of view, the dualist theories of life have had the object of showing us in a striking manner the interrelations among the three realms of nature. They have studied in particular the consequences of these relationships and have looked upon each being as a machine working in the service of others. These theories are especially imprinted with the finalist considerations that man cannot keep from expressing when he makes himself the center of the great cosmic phenomena that surround him: the mineral kingdom is the general reservoir; plants work for animals, and the entire world is made for man, who utilizes its products for his material well-being or in the interest of society. From this standpoint these theories seem to be related to the practicalities of everyday life. This is why numerous applications have been made of it in agriculture and hygiene which we do not need to examine here.

At any rate we believe that these theoretical views, which rest on evident and incontestable data do not correspond to a true physiological concept of the phenomena.

In fact, the identification of the animal organism with an apparatus in which active forces are engendered with a furnace into which the plant kingdom comes to be engulfed and burned, can represent outward appearances, but it is not the physiological expression of a law linking animal and plant life. Without doubts the herbivorous animals feed on plants, and the carnivors on the herbivors. These facts, which ensure the cosmic equilibrium, are the consequences, as we shall demonstrate later, of the general law of the struggle for existence, according to which nature cannot engender life except by death; creation by destruction. For us these facts, although necessary, are in reality accidental and contingent in their determinism; they remain outside of physiological finality.

The law of physiological finality is within each particular being and not outside of it; the living organism is made for itself, and it has its own intrinsic laws. It works for itself and not for others. There is nothing in the law of evolution of grass that implies that it should be cropped by a herbivor; nothing in the law of evolution of the herbivor that indicates that it must be devoured by a carnivor; nothing in the law of growth of cane that announces that its sugar must sweeten man's coffee. The sugar formed in the beet is not destined, either, to maintain the respiratory combustion of the animals that feed upon it; it is reserved for consumption by the beet itself in the second year of its growth, when it flowers and fructifies. The hen's egg is not laid to become the food of man, but rather to produce a chicken, etc. All these utilitarian ends at our disposal are works that belong to us (see Lecture VIII, final causes), and do not exist at all in nature outside of ourselves. Physiological law does not condemn living beings in advance to be eaten by others; animals and plants were created for living. On the other hand a compelling consequence of life is to be unable to be born except from death. We have repeated this in all its forms: organic creation implies organic destruction. What is observed in the intimate phenomena of nutrition, deep within our tissues, is manifested in the great cosmic phenomena of nature. Living beings cannot exist except with materials from other beings that have died before them or were destroyed by them. Such is the law.

In résumé, general physiology, which considers life only in its essential and general phenomena, does not permit us to accept a duality of animals and plants, or an animal physiology distinct from plant physiology. There is only one way of living, only one physiology for all living beings; it is general physiology that decided upon the vital unity of the two kingdoms.

If now, instead of considering life in its two necessary and universal manifestations, vital *creation* and *destruction,* we penetrate into the workings of the various vital mechanisms that nature reveals to us, if we descend into the arena where the struggle for existence takes place, then we shall find infinite functional differences and infinite varieties. Not only will we find that animals are adapted to eat plants, but that animals are equipped to devour other animals weaker than they. It is in a word the rule of the law of the strongest, a law that is in no way obligatory, since the hazards of the struggle for existence can permit one being to escape death, while another succumbs.

At any rate, in the course of this silent struggle, which we refer to antithetically as the harmony of nature, and into which all living beings come to mutually destroy themselves, the fundamental law of general physiology that we have enunciated is never violated. Life never manifests itself without being accompanied in the same individual by a double movement of equivalent organic creation and destruction, so that we never find living beings playing separately the role of organisms creating organic material while others have the opposite role of destroying this organic material and restoring it to the mineral world.

All living beings nourish themselves in the same way; neither the animal nor the plant proceeds by a direct nourishment; both, in reality and despite appearances to the contrary, nourish themselves by taking from the ambient world materials that have fallen into a more or less profound state of chemical indifference. The animal and the plant as well modify these materials, work on them, and from them form reserves appropriate to their nature, later utilizing them for their own benefit. At times the formation of the reserve and its expenditure can be almost simultaneous or very closely related; at times they follow one another and at a long interval. This last case is seen in plants, especially in biennial plants. The plant accumulates its reserves during the first year, and it might be thought that it is then only an apparatus of creation or of synthesis. In animals, on the contrary, and particularly in warm-blooded animals, the reserves do not last long, and are expended more or less measure by measure, so that it might be thought that these latter beings are solely an apparatus for combustion and destruction. In cold-blooded animals the reserves are in certain cases, made on a long-range basis, and in this way resemble those of plants.

After all, the plant and the animal are two distinct living machines, provided with various instruments and apparatus having ways of functioning that give very different appearances to the phenomena of their existence. But the unity of *life* ought not to be hidden from us by the variety of the *function;* the muscle, the gland, the brain, the nerves, the electric organs, etc., live in the same way, but they function very differently. Plants and animals live identically, but function otherwise. Even admitting that the function of chlorophyll is special to plants, the conclusion should not be drawn that plants live in some other way than animals; this would be an error; chlorophyllian protoplasm, having the function of reducing carbonic acid and giving off oxygen, does not live any less than all animal and plant protoplasms when absorbing oxygen and giving off carbonic acid.

From the point of view of general physiology, we do not consider only the functions that differentiate among the living beings; these present nothing absolutely necessary to life; we shall consider, on the contrary, the general and common phenomena that are indispensable to the existence of all beings. It matters little that one living being has more or less varied and complex organs or apparatus, lungs, heart, brain, glands, etc. etc. None of this is necessary for life in an absolute way. The lower beings live without these apparatus, that are only the appurtenances of the de luxe organizations. The study of the lower beings is especially useful in general physiology, because among them life exists in the naked state, so to speak. It is reduced to nutrition: vital destruction and creation. But, we repeat, this life is always complete; in the plant as well as in the animal. Neither one represents a half-life, which reciprocally completing each other, would make the two beings strictly complementary one to another.

It is eventually within the intimacy of the phenomena of nutrition that the law of vital unity is especially manifested in animals and plants. But to grasp this unity, it is necessary to consider the phenomenon of nutrition in its totality; for if we analyze only one aspect of the relationship between the living being and the cosmic environment, we can sometimes find that the phenomena of animal and plant life take on contrary appearances. It is this that at times appears to result from what is called the nutritional balance of animals and plants. We shall conclude with some reflections on this subject.

The balance of the organic exchanges in animals and plants is drawn up like that of an ordinary machine whose internal work one wishes to assess. What goes in is analyzed, what comes out within an allotted time is analyzed, and from the expenditure, what has taken place within the machine can be deduced. This procedure, doubtless applicable to inert machines, is no longer legitimate for living organisms or machines. If organic nutrition and combustion were *direct,* as it was believed after Lavoisier, a direct balance would be acceptable. But physiology has taught us that nutrition is *indirect,* and takes place only at long range over months and even years in certain plants. In order to come to any conclusions therefore, it would be necessary to have rigorous observations or experiments of an equal duration; otherwise partial results would be obtained, from which general conclusions could not be drawn.

Regnault and Reiset have well expressed this difference existing between living machines and inert machines when, in their excellent investigations on respiration, they analyzed the work of Dulong and Desprez on animal heat. These latter authors, implying that combustion is direct, admitted that the heat produced in the body is represented by the heat of combustion of carbon and hydrogen by means of the inhaled oxygen. The data from their analyses correspond, in fact, to this explanation. Regnault and Reiset, while admitting that the phenomena of calorification within the organism as well as outside of it cannot be anything but the result of phenomena of combustion, do not hesitate to consider the values found by Dulong and Desprez to be false, and the agreement of their analyses to be completely fortuitous. The fact is, there are a good many other phenomena that should be taken into account if one wants to work out the equation for the production of animal heat in the living organism.

These problems are therefore overly simplified, and according to the witty expression of Mulder, to deduce phenomena that take place in the organism by analyzing the materials that pass through it would be like attempting to ascertain what takes place in a house by analyzing the food that enter through the door and the smoke that comes out of the chimney.

We nevertheless recognize a great value in chemical balance studies, because they provide the primary facts as the base from which the physiologist can pursue the study of the intimate phenomena of nutrition in our tissues. But experimental physiology teaches us that these intermediary problems of nutrition must then be followed step by step with the help of delicate experiments, instead of being deduced from hypothetical explanations based on the comparison of materials on entry and exit.

The phenomena of nutrition are too complex to be able to lend themselves to this kind of investigation, which is applicable, we repeat, only to inorganic machines. We could cite many erroneous physiological conclusions, to which this indirect way of proceeding has led, while on the contrary, the experimental study of the phenomena of nutrition, pursued directly into the organs, tissues, and even into the elements of the tissues, has led us to fruitful discoveries. The formation of sugar in the liver would never have been discovered if analysis had been limited to a comparison of materials entering and leaving the organism. The physiologist must base himself on these chemical facts, but he must not be content with this; he must, by means of direct experiment, get down into the intimacy of the organs, tissues, and into the living cell, whose function is identical in the animal as well as in the plant. It is only by such a study that he will be able to grasp the mystery of intimate nutrition, and succeed in mastering these phenomena of life, which is his supreme goal.

It can thus be seen how the point of view of the physiologist and the chemist can differ when they study the phenomena of the living organism.

CONCLUSION

From the preceding general discussion, we can conclude that despite the genuine variety that vital phenomena present to us in their external appearance, in animals and in plants, they are fundamentally identical, because the nutrition of plant and animal cells, which are the only essential living parts, could not have different modes of existing in the two realms.

Consequently, we consider our great division of the phenomena of life, *organic destruction* and *creation,* as justified and established in general physiology. This division will serve us as a framework in the lectures to follow.

FOURTH LECTURE

Phenomena of Organic Destruction

Fermentation—Combustion—Putrefaction

SUMMARY: Phenomena of organic creation and destruction—Study of the phenomena of organic destruction: Fermentation, combustion, putrefaction.

I. Fermentation—Catalysis; Berzélius—Decomposition; Liebig—Organic theory; Cagniard de Latour, Turpin, Pasteur—Soluble ferments, organized ferments—Actions of soluble ferments found in the mineral kingdom—The same ferments are common to the two kingdoms, animal and plant—The ferments act to transform and decompose the products of the nutritional reserves—Fermentation due to organized ferments—Alcoholic fermentation; its conditions.

II. Combustion—Theory of Lavoisier; direct, active, or latent combustion—Direct combustion does not exist—Indirect combustion, cleavage, kind of fermentation appertaining to plants and animals—Particular case of glands—Unknown role of oxygen in the organism.

III. Putrefaction—Appertains to animals and to plants—Theories of putrefaction; Gay-Lussac, Appert, Schwann, Pasteur—Putrid fermentation—Analogy between putrefaction and fermentation—Life is a putrefaction—Mitscherlich, Hoppe-Seyler, Schutzenberger, etc.

We have proposed, discussed, and established in general physiology the division of the phenomena of life into two major groups: *phenomena of creation or organic synthesis,* and *phenom-*

ena of organic destruction. It is now necessary to pursue this division in detail and study separately the two orders of vital phenomena referred to therein. We shall begin with the study of the phenomena of vital destruction, because they can be seen in the organism from the beginning, and make their debut with the first appearance of life.

The phenomena of organic destruction have the vital manifestations as their expression. The following proposition can be regarded as an axiom in physiology:

Every manifestation of life is necessarily associated with an organic destruction.

What are the phenomena of this disorganization?

In the passage we quoted above Lavoisier related all the phenomena of organic destruction to one of these three types:

I. Fermentation.
II. Combustion.
III. Putrefaction.

It is, in fact, by one or the other of these processes that organized material is destroyed, either as a consequence of vital activity, or in the cadaver after death. Unfortunately, these three typical phenomena still present many obscurities, despite the very active impetus that has been given to their study and despite the considerable progress that has been made in the last several years In these lectures, in which we are outlining a sort of sketch or plan of general physiology, there is no question, for that matter, of resolving these problems; it is first necessary to state them. We shall restrict ourselves to this in treating, in succession, fermentation, combustion, and putrefaction. We shall point out in a rapid and summary fashion, not the detailed state of our knowledge about these complex phenomena, but rather the place that they ought to occupy in a general survey of physiology, reserving for a later time their development by communication of our personal investigations.

1. FERMENTATION

Chemists and physiologists have never been, and still are not in agreement about what ought to be meant by the word fermentation. In recent times it has been said that in general this name was applicable to all organic reactions induced by a material that gained nothing and lost nothing in the process, but which seemed to intervene only by its presence. Berzelius called phenomena of this kind *catalytic actions*. So it is that platinum sponge, it is said, acts by its *mere presence* or by catalysis on alcohol to make it pass successively into the state of aldehyde, and then into acetic acid. Fermentation would thus be an organic catalysis. This would be, of course, a simple description and not an explanation. The relationship that this name implies is not accurate, however, and would give us a most incorrect idea of the fermentations that take place in animals and plants.

In fact, the fermentations that we know from having studied them in the living economy where they take place are not to be compared with the phenomena that Berzelius called *catalytic actions*. The ferment does not remain indifferent to the decompositions that it produces. It is now proven that in the action of diastase on starch the diastase is used up, and its depletion is in proportion to the intensity of the action it has exerted.

Thus the ferment does not remain unchanged. We have just cited a case in which it is destroyed; in other cases it multiplies. This takes place in what are called the organized ferments. *Mycoderma aceti,* the microscopic organism that transforms alcohol into acetic acid, does not act simply in the manner of platinum sponge; it increases in weight, it grows and multiplies in the fluid in which it acts, in proportion to its own activity.

According to this, fermentation should not then be compared with otherwise obscure and unknown phenomena that have been classified under the heading of *catalytic actions*. Berzelius had alcoholic fermentation especially in mind; he did not know that the ferment yeast was an organized being; he regarded it as an amorphous principle. Mitscherlich, although he knew of the organized nature of yeast, attributed to it the same role as Berzelius.

Liebig understood fermentations differently. Taking alcoholic fermentation as the prototype, he considered it as the iatrochemists Willis and Stahl had done in the past. "Brewer's yeast, and in general all animal and plant materials in the course of putrefaction transfer to other bodies the state of decomposition in which they find themselves; by disturbing the equilibrium, the activity to which their own elements are exposed is communicated equally to the elements of bodies which find themselves in contact with them." According to this viewpoint, the ferment is a decomposing body whose molecules, animated by a particular internal activity, communicate the disturbance to an unstable fermentable substance.

To characterize Liebig's theory in a word, it could be said that fermentation is a decomposition that sets off another.

Around 1836, Cagniard de Latour recognized by microscopic inspection that the yeast of alcoholic fermentation was formed of organized globules of living cells, capable of reproducing, having a membrane and internal contents. It was Pasteur above all who defined most precisely the role of this organism in fermentation. Alcoholic fermentation is a phenomenon related to the organization, the development, and the multiplication, that is to say to the life of the globules. This is what has been called the physiological theory of fermentation, which Turpin in 1838 was the first to formulate by saying "Fermentation as effect and vegetation as cause."

Today, two kinds of fermentations are distinguished, according to the soluble or insoluble nature of the ferment; the one produced by the intervention of an organized or *structured* ferment, the other produced by nonorganized ferments, liquids, *soluble* products, that are elaborated and secreted by the living organisms.

Soluble ferments are found in plants and animals. They have as prototypes plant diastase and the digestive ferments; they have as a common characteristic, solubility in water, precipitation by alcohol, and resolubility in water. Another common feature is the magnitude of the effect compared with the very small mass of the ferment. A very small proportion of diastase can turn a great quantity (more than two thousand times its own weight) of starch into sugar. Finally the active substance does not multiply, but on the contrary is depleted and destroyed by its own action.

These ferments are capable of producing very vigorous chemical reactions. I have tried for a very long time to establish that fermentations, however special in their processes, are not basically, in their essential nature, different from general chemical actions; all in fact are represented in the mineral kingdom. Certain ferments, such as animal and plant *diastase,* and inverting ferments in plants or animals, act like mineral acids; others have the same effect that alkalis produce; among these is the ferment of fatty material that is found in pancreatic juice which first emulsifies and then saponifies these substances, etc.

Fermentation brings about the destruction of complex compounds in organisms, splitting them into simpler bodies, accompanied by hydration. They play a very important role in nutrition. They are found in the economy of plants as well as animals. The fact is easy to demonstrate in the case of the diastases; the glycolytic ferment, or diastase properly so called is found in all parts of the organism where animal or plant starch must be made soluble. In seeds the ferment manifests its activity at the time of germination; in the tuber of the potato it becomes active in the spring; in the liver it is always present in order to transform animal starch into glucose. In other words, wherever starchy materials must nourish an organism, the presence of an identical ferment can be established. Starch is therefore not utilized in its actual form; it participates in the life of the plant or animal only when, by hydration, it has been converted into the sugar glucose. Moreover, if the sugar were in the state of glucose, it would not remain in the organism; it would soon be destroyed without being able to play that role of reserve that is indispensable to vital functions within the two kingdoms.

What we say about starch, its accumulation in insoluble reserves, and its conversion by fermentation at the appropriate time is true for many other substances not so well known. The behavior of one of these, however, the sugar saccharose (cane or beet sugar) serves to confirm this generalization. It is able, in fact, to accumulate as a reserve in plant tissues. In this form it is not utilizable; it is not directly oxidized by the organism; it is necessary for it to be converted into the sugar glucose. An *inverting ferment* is responsible for the transformation. This ferment exists in identical form in animals and plants: brewer's yeast, which must convert the cane sugar with which it is placed in contact into glucose for its nourishment, manufactures this ferment. Berthelot discovered it there. The beet behaves in the same way with regard to the sugar accumulated in its root during the first year of growth, and I have demonstrated that animals proceed in the same way to make use of the saccharose sugar contained in their food.

We have said that actions of the type of fermentation are extremely numerous; they are in fact the general type of the vital actions of destruction; many are as yet only suspected; the greatest number are absolutely unknown. What is known is enough to make it possible to estimate the importance of these phenomena.

Albuminoid materials are rendered soluble and are digested by a ferment, *pepsin,* which exists in the gastric juice; pepsin only begins the action, *trypsin,* a ferment of the same nature contained in the pancreatic juice completes this conversion into peptone. It has been thought that this agent exists in the different parts of the organism wherever its presence would be necessary for the digestion of albuminoids; Brücke has claimed to have found it in blood and muscle. It is probable that it will be isolated from plants.

Similarly, there exists in almonds, sweet and bitter, an active soluble ferment, *emulsin,* which is capable of splitting a great number of glycosides: amygdalin (into glucose, hydrocyanic acid, and essence of bitter almonds), salycin, helicin, arbutin, phlorizin, aesculin, and daphnin. Now it is remarkable that a ferment of the same nature is found precisely in animals, in the liver, and the pancreas. It would be useless to multiply these examples, to call attention to the fermentation of myronate of potassium by myrosin and the fermentation of the bile acids, of hippuric acid, of tannin, of pectin, etc. It is enough to understand that here is a general process employed by nature to promote the cleavage, that is to say, the destruction, of a great number of organic principles in plants as well as animals.

A second kind of decomposition produced by organized beings is included among the fermentations (F. by *organized ferments).* The prototype of these actions is the alcoholic fermentation produced by brewer's yeast.

Among this group of phenomena there should be included

the conversion of sugar into alcohol, into lactic acid, into butyric acid, into gum, into mannose, and into acetic acid.

These are examples of *destruction* brought about under special circumstances or in the course of the existence of particular organisms.

Nevertheless, some of these destructive fermentations of organized materials might perhaps have a very wide distribution. It would seem that many cells, either animal or plant, when placed in the same conditions as yeast cells, react like them.

Under what circumstances does yeast provoke alcoholic fermentation? According to Pasteur it is when the ferment is deprived of air. Since it needs oxygen to survive, and not being able to obtain it directly, it is faced with the alternative of perishing or of procuring it in some way. In these circumstances the yeast takes oxygen from the surrounding materials; it takes it from sugar by producing its fermentation or destruction, an operation capable of engendering heat, and of producing the thermal energy expended in the vital function.

It is known, as we have said, that other cells seem to be capable of acting in an identical manner. It has been pointed out in fact that certain African plants produce alcohol in their roots. Lechartier and Bellamy have shown that fruits placed in an atmosphere of carbonic acid, that is to say, rendered incapable of breathing as they ordinarily do by absorbing oxygen and giving off carbonic acid, behave like yeast; they partially transform their sugar into alcohol and carbonic acid. It is known moreover that it is possible to extract alcohol by the distillation of certain fruits, such as prunes, when they are ripe. de Luca was convinced that certain leaves placed similarly in an atmosphere of carbonic acid behave in the same way and give rise to alcoholic and acetic fermentations.

The fermentation caused by organized or living ferments might be compared to a sort of parasitism that alters the environment in which these elementary beings live. For this reason these ferments are included in our study, since they cause the destruction, or cleavage, of the more simple materials with which they are in contact.

II. COMBUSTION

We have no intention of entering upon the study of the phenomena of combustion and of their role in the life of organisms. We wish only to recall, on this occasion, a principle that we have maintained for a long time, namely that chemical phenomena in living organisms can never be fully equated with phenomena that take place outside them. This means to say, in other words, that the chemical phenomena of living beings, although they take place according to the general laws of chemistry, always have their own special apparatus and processes (See, on this subject, my *Rapport sur les Progrès de la Physiologie Générale,* 1867).

It has been known since Lavoisier that the destruction, the molecular wear and tear, that accompanies vital phenomena consists of a sort of oxidation of the organic material; it is the equivalent of a combustion. But Lavoisier and the chemists who made this important fact known to us, fell into a certain error, nearly inevitable at the time, concerning the mechanisms of these phenomena, an error that even today is still in vogue among many scientists. They equated the chemical processes that take place in the organism to a direct oxidation, to a fixation of oxygen upon the carbon within the tissues. In a word, they believed that organic combustion had as its prototype the combustion that takes place outside living beings, in our fireplaces and in our laboratories. Quite to the contrary there is within the organism perhaps not a single one of these phenomena of this alleged combustion that takes place by the direct fixation of oxygen. All of them make use of the services of special agents, of the ferments, for example.

The immortal work of Lavoisier on respiration gave us an understanding of the role of oxygen, if not in its details, but at least in its broad outlines. Oxygen is necessary for the maintenance of life, it has been said, because it supports combustion; its suppression, if it is not compensated by some artifice, cannot be sustained for long; this gas unites with the organic substance and is eliminated from the organism in a state of combination with carbon, as carbonic acid.

It is not however in direct combustion that this gas is utilized. The common formula repeated by all physiologists that the role of oxygen is to support combustion is not accurate, since in reality there is no true combustion within the organism. What is true is that the exact role of oxygen, which we believe we know, is still unknown to us; it can hardly be suspected. Here we can only state the question, without pretending in any way to resolve it, but in any event we already know that oxygen does not serve for direct combustion.

First of all, what do the chemists mean by the word *combustion?* It is still one of those imprecise terms about which the most complete disagreement exists. Some chemists reserve the term for the oxidation of carbon and hydrogen, having as a consequence the production of carbonic acid and water vapor, with the production of heat. With Lavoisier they distinguish between active combustion and slow combustion, according to whether the production of heat is more or less intense, whether it is dissipated as it is formed so as not to raise the combustible body to a high temperature in the case of slow combustion, or whether it is on the contrary carried to such a degree that the body becomes incandescent in the case of active combustion.

Other chemists consider the characteristic feature of combustion to be the development of heat, so that they give this name to every combination, to every chemical reaction which is accompanied by a great development of heat.

While adhering for ourselves to the first meaning, can it be said that there is combustion in the animal or plant organism? The response to this question has been in the affirmative.

Lavoisier, who, with the intuition of genius created his system by comparing respiratory phenomena to the oxidation of metals, had to believe that this was so. He had compared (1789) the consumption of oxygen in the same person, first at rest, and then while doing work, and had concluded that muscular work accelerated organic combustion. Since then, everyone has been so fully convinced that there is a genuine combustion, that the debate centered simply on the question of ascertaining whether it was the substance of the muscle itself that burned, or whether it was combustible hydrocarbonaceous materials.

But neither of these opinions could be maintained insofar as they implied a direct combustion. In fact, products of incomplete combustion such as carbon monoxide are never encountered within the organism. Moreover, there is no burning of hydrogen; it has never been possible to establish directly the production of water in organic combustions. It seems to the contrary to be well established that the water of the organism has its source exclusively in the food, and that it is introduced from without. I have shown that the blood that leaves contracting muscle is not richer in water than that which enters: it is often even the opposite. I have moreover pointed out that the blood that leaves a gland during secretion is poorer in water than that which enters, and that the difference is represented exactly by the quantity of water contained in the fluid secreted.

Oxygen is, moreover, not employed immediately; it is not directly fixed. An active muscle produces a quantity of carbonic acid greater than the quantity of oxygen absorbed at the same time. The consumption of oxygen is thus not in exact proportion with the production of carbonic acid. This is what Petenkofer and Voit have established for the muscle *in situ,* and for the muscle removed from the animal, L. Hermann has obtained the same result. It is known (and we shall repeat the experiment here before your eyes) that even in the absence of any renewal of oxygen, in inert gases such as hydrogen, for example, which we have substituted for ordinary air, the muscle can contract for quite a long time. It still gives off carbonic acid, which obviously does not come from direct combustion. If during activity the muscle gives off more combined oxygen than it receives, on the contrary, during rest it takes up more of it than it gives off. The facts establish quite clearly that here there is no question at all of a direct and immediate fixation of oxygen upon the substance of the muscle. The phenomenon is much more complex. It consists in chemical cleavages, most certainly of the nature of fermentations, but at present rather suspected than understood. The hypothesis has been invented of a cleavage by fermentation of a muscular substance, *inogen,* into *carbonic acid, sarcolactic acid,* and *myosin.* This hypothesis is valuable simply to show us the direction of the present-day interpretations that tend to be substituted for Lavoisier's theory of direct combustion.

Study of the functional activity of the glands leads to conclusions of the same nature regarding direct combustion. I have shown that the venous blood leaving the glands is nearly as rich in oxygen as the arterial blood, so that an increase in function does not entail the disappearance of oxygen. Thus, oxygen is not fixed at the time when it might be supposed that it ought to be used; in a word, there is not a greater consumption of oxygen. Yet, it is during functional activity that the greatest quantity of carbonic acid is produced, which is found in considerable quantities in the crimson venous blood, loaded at the same time with oxygen and carbonic acid. Thus, the two phenomena of absorption and utilization of oxygen are here clearly separated, which obviously excludes any possibility of direct combustion. It is during rest that oxygen is absorbed by the gland; it is during functional activity that it leaves in the form of carbonic acid, but then the absorption of oxygen is suspended.

It results from these facts that oxygen is not employed in direct combustion, an important consequence for the goal we pursue, for the direct combustion of carbon and hydrogen would be a true synthesis, a combination of separate elements, while on the contrary the phenomenon that occurs is probably a cleavage, a destruction of a complex substance, or a true degradation by fermentation.

We have said above that the true role of oxygen is unknown. It is indeed certain that this gas is fixed within the organism and that in this way it becomes one of the elements of the constitution or of the organic creation. But it is by no means by its combination with organic material that it provokes vital activity. By entering into contact with the parts, it makes them excitable; they can live only upon the condition of this contact. It is thus as an agent of excitation that it intervenes immediately in the majority of the phenomena of life.

It has been said that in the higher animals oxygen must be carried to the nervous system to excite the medulla oblongata and generate the respiratory movements. In the frog the necessity for excitability is less in the winter, the period of inertia, than during the summer, the period of activity. Thus the absorption of oxygen is less during the former season than during the latter. A singular experiment by Engelmann seems to throw some light on this role of excitant that oxygen plays. Engelmann observed the movements of the vibratile cilia, movements that are easy to observe after the membrane that carries them has been detached from the animal. The vibratile cells are examined under the microscope. If the oxygen is driven off the preparation and replaced by hydrogen, the movements cease after a certain time, about twenty minutes for example. If oxygen is readmitted the movements are resumed and these alternations can be repeated a certain number of times. Oxygen acts therefore as though it excited the vibratile motions, and as though its power of excitation continued for a certain time. If vibratile cells, slowed down by the cold and winter dormancy, are taken and the experiment is repeated, the same results are obtained, only the action of oxygen continues for a greater length of time, is effective for a longer period, and motions still continue for several hours after contact with the gas.

The conclusion we put forward at the beginning seems to us to be amply justified; therefore, it is not necessary to multiply examples further to prove that the theory of direct combustion, which led to such great progress when its illustrious founder introduced it to science, has nevertheless not been confirmed by physiological studies. Combustion is not direct in the organisms, and the production of carbonic acid which is such a general phenomenon in vital manifestations is the result of a true organic destruction, of a cleavage similar to those produced by the fermentations. These fermentations are moreover the dynamic equivalent of combustion; they fulfill the same purposes in the sense that they engender heat and are consequently a source of the energy that is necessary for life.

III. PUTREFACTION

Among the processes of destruction of organic materials, Lavoisier placed *putrefaction* along with fermentation and combustion. This concerns a phenomenon still more obscure than those of fermentation and combustion which we have examined previously.

What is meant by putrefaction? It has been known at all times that the materials that enter into the composition of the animal body start to alter after death, to be transformed and decomposed into various principles, including substances with a strong and putrid odor. From this comes the name putrefaction, to characterize those decompositions with a nauseating odor.

The same thing takes place in plants. Only here, since the destruction acts on bodies in which the albuminoid or nitrogenous substances are in lesser quantity, the organoleptic characteristics of putrefaction are less striking, and have been less well known. In reality, the substances of the plant organism, the active working and truly living substances such as albuminoid protoplasm, are just as liable to putrefaction as in animals. Only, as we have just said, the proportion of living parts in the individual plant is very small in comparison to the inert supporting or skeletal parts. These are no more susceptible to putrefaction in animals than in plants; the carapace of a crustacean, the skeleton of a mammal are in an unalterable state similar to the bark or the wood of an oak.

After the work of Appert and of Gay-Lussac, it was thought that putrefaction was a decomposition, a cleavage produced by the momentary intervention of oxygen, continuing from then on by a sort of communicated molecular movement.

Later, the work of Schwann, Ure, Helmholtz, and especially of Pasteur showed that the determining cause of putrefactions must be sought in the microscopic beings, the vibrios, bacteria, and molds that develop in decomposing liquids, whatever opinion might otherwise be formed of the origin of these beings. Alterable substances lose this property when all the air is driven off by boiling, and when nothing is permitted to enter the vessel that contains them but air previously heated red-hot.

Pasteur distinguished two kinds of putrefactions; those that take place in the absence of oxygen which he has called *putrid fermentations,* and those in which oxygen intervenes as an essential element the one and the other however being produced by organisms.

Putrid fermentation will manifest itself in a liquid when it no longer contains oxygen, when the first infusoria to develop have consumed it all. Then the "vibrio ferments, which have no need for this gas to live, begin to appear, and putrefaction at once declares itself. It accelerates little by little, following the progressive advance of the development of the vibrios. As to the putridity, it becomes so intense that microscopic examination of a single drop of fluid under the microscope is a painful matter."

The products of putrefaction are very numerous; each albuminoid substance can, so to speak, behave differently in this regard. There are, as nearly constant ternary compounds, volatile fatty acids, simple and complex amines, leucine, tyrosine, carbonic acid, hydrogen sulfide, hydrogen, and nitrogen.

The second kind of putrefactions includes those that require the presence of the oxygen of the air; these actions, called *putrefaction, slow combustion, eremacausis,* destroy animal or plant organic materials exposed to the air, and after more or less complex transformations reduce them to carbonic acid, water, nitrogen, and ammonia which return to the atmosphere.

According to Pasteur, these actions are again due to organisms, molds, and bacterias; these slow and spontaneous combustions would never take place without the development of organisms within or upon the surface of the substances that decay.

In ordinary circumstances the two kinds of actions take place simultaneously or successively. An alterable substance being left in the air, the oxygen is first taken away by the first infusorias to appear *(Monas crepusculum* and *Bacterium termo).* The fluid becomes cloudy. A pellicle forms on the surface, impeding the access of air; the putrid fermentation of the vibrios takes place in this anoxic fluid. The pellicle falls to the bottom. New bacteria form on the surface and produce putrefaction or slow combustion; but then the same cycle of events begins again and continues until the alterable material is completely exhausted.

This is where our understanding of putrefaction stands today. Are there actions of this kind, identical in their processes, that can take place in the living organism and destroy the organic material in them?

The organism does not normally permit the development or the introduction into its interior of these parasitic bacteria and these vibrios. Nevertheless it is possible under certain circumstances that phenomena of the same nature really take place.

Chemists who are skilled and expert in studies of this kind do not fear to support this. It was a long time ago that I heard

Mitscherlich say "life is nothing but a putrefaction." Hoppe-Seyler (1875) expressed himself similarly somewhere, "without wishing to propose in principle the identity of organic life with putrefaction, I shall nevertheless say that to me the vital phenomena of plants and animals have no more perfect analogy in all of nature than the putrefactions."

It can be thus accepted that in organisms there can exist processes analogous to those of putrefaction. The organic substances would undergo the same transformations and the same cleavages that take place in putrefaction.

What is peculiar to the mechanisms of putrefaction? Looking at the question from the chemical point of view, it could be said with Hoppe-Seyler that the essential fact is a modification of the molecular equilibrium of the substance with transport of the oxygen from the hydrogen to the carbon atom, this action being expressed in some cases by the liberation of carbonic acid accompanied by the elimination of hydrogen or more fully hydrogenated compounds. All the other phenomena that take place are preceded and conditioned by this one; they are secondary phenomena produced by the hydrogen in the nascent state, or by the purely chemical and subsequent intervention of the oxygen contained in the medium.

It is phenomena of this kind that would be produced by the organisms reported by Pasteur, the lactic ferment, the butyric ferment, etc. But it could happen, as has already been demonstrated regarding the alcoholic fermentation of yeast, that other cells and elements of the organism behave in the same way. In fact, all the chemical changes within the organism would be included in this kind of theoretical action, and this is a theory that might be proposed to substitute as a hypothesis for the demonstrably false hypothesis of direct oxidation.

Putrefactions are moreover characterized by phenomena of cleavage with final products that have been well studied by Schützenberger. I have seen that of all the organs of the body, the one that putrifies most easily is the pancreas. A particular and final characteristic of this putrefaction is a red coloration, first observed by Tiedemann and Gmelin. I studied it next, and recently in my laboratory Prat established that this red material manifests itself in the putrefaction of nearly all nitrogenous substances, animal or plant. This red coloration that Prat is studying at the moment might be due to a poorly known product of putrefaction.

CONCLUSION

Without caring to enter any further into the question of the organic decompositions, which is still surrounded by the greatest obscurity, we shall limit ourselves to drawing a single general conclusion from this lecture:

> Putrefaction like combustion is related to fermentation. All the actions of organic decomposition, or of vital destruction, that take place within the organism amount, in sum, to fermentations. Fermentation would be the general chemical process for all living beings, and would even be peculiar to them, since it does not take place outside of them. Thus fermentation characterizes living chemistry, and consequently its study belongs strictly within the domain of physiology.

FIFTH LECTURE

Phenomena of Organic Creation

Anatomical Theories: Cellular, Protoplasmic, Plastidular

SUMMARY: Organic creation, comprising two orders of phenomena common to the two realms: chemical synthesis, morphological synthesis.

I. Anatomical Constitution and Morphological Creation of the Living Being, Animal or Plant; History—Early period: Galen, Morgagni, Fallopius, Pinel, Bichat, Mayer—Modern period: de Mirbel, R. Brown, Schleiden, Schwann—Cell Theory—The ultimate morphological element of living beings is the cell, but a living substance antedates the cell; this is the protoplasm—It is the seat of chemical syntheses, of morphological syntheses.

II. Origin of the Cell in Protoplasm—Protoplasmic theory—Blastema—Gymnocytode, Lepocytode—Protoplasm in plant cells—The primordial utricle—The protoplasm is the living body of the cell in the two kingdoms.

III. Protoplasm; its constitution—Protoplasmic mass, nucleus—Protoplasmic beings—Monera, Bathybius—Structure of the protoplasm—Plastidular Theory—Complexity of protoplasm—Its role in the division of the nucleus—Relationships of nucleus and protoplasm—The nucleolus, its constitution, its role—Conclusion.

At the same time that the animal or plant organism destroys itself by the very fact of vital activity, it restores itself by a sort of organizing synthesis, a formative process that we have called *vital creation* and which constitutes the counterpart of *vital destruction*.

The act of vital repair does not have the same activity in all parts of the body however. There are parts in animals and in plants that are more vital, more delicate, and more destructible, while others, more resistant and of more obscure vitality, leave lasting traces of their existence after the death of the organisms. Such are the lignin or bones, that constitute the skeleton of plant and animal beings.

The act of synthesis by which the organism thus maintains itself is basically of the same nature as that by which it constitutes itself within the egg. This act is also similar to the process by which the organism repairs itself when it has undergone some sort of mutilation. Procreation, regeneration, redintegration, and cicatrization are the various aspects of an identical phenomenon, i.e. organizing synthesis or organic creation.

This organic creation exists at two levels. At times it assimilates material from the environment to form it into organic principles destined to be destroyed at a later time; at times it forms the elements of the tissues directly. Thus the formation of the immediate principles that constitute the reserves, that *pabulum* of life, that is to say *chemical synthesis,* must be distinguished from the combination of these principles into a particular mould, into a definite form or shape, according to the plan or design of the individual, of the tissues that form it, and of the elements of these tissues, that is to say, *morphological synthesis.*

We shall treat these two questions one after the other, first examining how the anatomists have succeeded, by gradually analyzing the living organism, in reducing it to its elementary parts; then we shall see how the physiologists and chemists have accounted for their synthetic creation.

HISTORY

The constitution of organisms has been studied from the beginning of the life sciences. In them were found the elementary parts of the organs and then of the tissues. In antiquity Galen had tried to analyze the organism into like parts.

Much later Morgagni attempted a grouping of the same kind, not only for healthy parts but also for those that were diseased.

Fallopius (1523-1562) assembled like parts into ten or eleven groups: bones, cartilages, nerves, tendons, aponeuroses, membranes, arteries, veins, fat, and bone marrow.

Finally, Pinel, the immediate predecessor of Bichat, opened the way for him by combining (according to still very incomplete pathological considerations) the anatomical parts that he considered to be analogous, for example the *hyaline membranes,* periosteum, dura mater, ligamentous capsules, pleura, peritoneum, and pericardium. But it was Bichat who had the glory of setting out masterfully along this road, so timidly opened. And, a remarkable fact that well illustrates the influence of precursors on the development of even the most original geniuses, it was through a critique of the classification of the membranes by Pinel that Bichat began his work in general anatomy.

In place of the descriptive anatomy cultivated until then which revealed the organism by describing its different parts in topographical order *de capite ad calcem,* Bichat instituted an infinitely more philosophical method, by collecting similar organs, however diversely located into the same group, and studying them together under the designation of systems: osseous, glandular, nervous, serous systems, etc.

In this analysis he employed, not the optical instruments which he rejected, and which have been such a great resource to his successors, but the much more imperfect method of dissociation, maceration, and the various chemical agents that permit a more detailed dissection. He nevertheless succeeded in laying the foundations for the science of living tissues. Says Bichat:

> All animals are an assemblage of various organs which, each executing a function, contributes in its own way to the preservation of the whole. They are like just so many specialized machines within the total machine that constitutes the individual. But these special machines are themselves made up of various tissues of very different natures, which in reality form the elements of these organs.

Bichat distinguished 21 kinds of tissues, which are found with their characteristics in the various parts of the same animal or in the same parts of various animals. From this the name of *General Anatomy* was given to their study. These 21 tissues were:

1. cellular tissue
2. nervous tissue of animal life
3. nervous tissue of organic life
4. arterial tissue
5. venous tissue
6. tissue of the exhalant vessels
7. tissue of the lymphatic vessels and glands
8. bones
9. bone-marrow
10. cartilages
11. fibrous tissue
12. fibrocartilagenous tissue
13. muscles of animal life
14. muscles of organic life
15. mucosa
16. serosa
17. synovia
18. glands
19. dermis
20. epidermis
21. hair

To each of these tissues he attributed special properties which are the physiological causes of the phenomena each exhibits To Bichat's mind physiology would be no more than the study of these vital properties, just as physics is the study of the physical properties of inanimate matter.

The fundamentals of the science created by Bichat were rapidly expanded and research was perfected thanks to the use of a very powerful analytic instrument, the microscope. The first *simple microscope* had been made in 1590 by the Dutchman L. Jansen. Malpighi (1628-1694) and Leeuwenhoeck (1632-1725) made great use of this instrument, to which they owed remarkable discoveries. Schwammerdamm (1630-1685) and Ruysch (1638-1731) did not understand the importance of the revolution that the use of this valuable instrument would bring.

The simple microscope was, however, inconvenient and inadequate; the compound microscope as used today was not to be developed until after Bichat, from 1807 to 1811, thanks to Van Deyl and Frauenhofer.

Bichat's studies thus marked the first step in the analysis of the composition of organisms. But life was still to be decentralized beyond the limits he assigned it, beyond the tissues. Life resides, in fact, not only in the tissues, but in the formed elements of these tissues, and even more profoundly in the formless substratum of these elements themselves; in the protoplasm.

In 1819 Mayer began to classify the elements of the tissues;

for the first time he used the name *histology,* an inappropriate term however, which has served to designate the new science.

I. CELL THEORY

From this moment one begins to be concerned not only with knowing the elements of the various tissues, but in addition with uncovering their origin, with ascertaining their provenance, in a word, with *histogenesis.*

Studying plants, Mirbel reports that they all come from an identical tissue, the cellular tissue; that they have the cell as their elementary component. R. Brown discovers the nucleus of the cell.

The works of Schleiden and Schwann were to found the *Cell Theory.* Th. Schwann in 1839 showed that all the elements of the organism, whatever its actual state, have had a cell as a starting point. Schleiden provided the same demonstration for the plant kingdom, so that the origin of all living beings could be traced back to this simple little organism, *the cell.*

The *cell,* therefore, is the *anatomical element* of plants and animals, the simplest morphological organism of which the complex beings are constituted. There are plants that are constituted solely of cells (cellular tissue, parenchyma). At other times the cells are united into *vessels* or are transformed into *fibers.* The most complex plant is a combination of vessels, of fibers, of cells, that is to say, in sum, of more or less modified cells.

What we have just seen in regard to plants is true of animals. The elements of all tissues have been relegated by the histologists to the form of cells. In addition to the well characterized cells come the blood corpuscles, hematocytes, and leucocytes, the fusiform bodies of embryonic connective tissues, stellate pigmented bodies, the elements of the hepatic gland, the smooth fibers, and the myeloplaxes, which are cells in different anatomical states. It was recognized (Remak, 1852; Max Schultze, 1861) that the element of voluntary muscle, the striated fiber, develops at the expense of a single cell whose nucleus divided or proliferated. Still more recently, my former collaborator, Ranvier, now Professor at the Collège de France, has related to a cellular prototype an element which had seemed to be an exception, the nerve fiber. He showed that a nerve fiber is composed of units placed end to end, true cells, whose considerable length (1 millimeter in adult mammals) had prevented recognition under the microscope until then.

In résumé, it is now established in a general way, thanks to the accumulated work of histologists, that the organism is constituted by an assemblage of cells, more or less recognizable, modified in various degrees, and associated and assembled in various ways. Thus for the 21 elements of Bichat, the 21 tissues that for him formed the materials of the organism, we have substituted a single element, the cell, identical in the two realms, in animals as well as in plants, a feature that demonstrates the unity of structure of all living beings.

The egg itself would be no more than a cell. In a word, the cell would be the primary representative of life It is thus to this element, the cell, that we ought now to associate the phenomena of creation and of organic synthesis, in the plant kingdom as well as in the animal kingdom.

As to the origin of this cell, of this body from which the organism starts, this has been interpreted in two different ways. Schwann, founder of the cell theory, accepted that cells could be formed independently of already existing cells by spontaneous generation, or better, by a sort of crystallization in an appropriate medium, the *blastema.*

"There is found," he says, "either in cells already in existence, or between cells, a substance without a definite texture, corresponding to the cell content, or to the intercellular substance. This material, or *cytoblastema,* has, thanks to its chemical composition and degree of vitality, the capacity of giving rise to new cells." Gerlach was one of the strongest adherents to this theory. Ch. Robin* in France, advanced similar views.

This theory survived without contradiction until 1852, when Remak showed that in the development of the embryo the new cells that appear are always derived from an earlier cell. In this the analogy is complete with plant tissues where the new elements

*Robin, *Anatomie et Physiologie Cellulaires.* Paris, 1873.

always have antecedents of the same form. Virchow* completed the demonstration by examining cellular proliferation in pathological cases. Thus, in opposition to the theory of the blastema or the equivocal generation of cells, there developed the cell theory that can be formulated in the adage, *"omnis cellula e cellula."*

II. PROTOPLASMIC THEORY

Science has not completely accepted this conclusion; it has been recognized that life begins before the cell. The cell is already a complex organism.

There is a living substance, the protoplasm, which gives rise to the cell and which is antecedent to it.

The *Cell Theory,* born in 1838 as a result of the work of the botanist Schleiden, began to be upset around 1850. At that time the plasmatic or protoplasmic theory made its appearance. It was again a botanist P. Cohn, who traced its first outlines. This anatomist observed the zoospores and the antherozoids of the algae, elements simpler than the cell in the sense that they are formed by a mass of protoplasmic substance, naked and without an envelope.

This notion of elements without an envelope passed at once into the animal kingdom. In 1850 Remak established that the first embryonic cells arising from the segmentation of the egg have no envelope at all, but are composed only of a mass of substance within which a nucleus is found.

In 1861 Max Schultze included in this category elements which at first sight had appeared to be unrelated to it, namely muscle fibers. He regarded as individual elements the bodies that are still called *nuclei* in the *muscle fiber,* because he found around them a fine layer of protoplasm; the same interpretation was soon thereafter extended to nerve cells.

The ultimate element in which life is incorporated is in that case no longer a *cell,* it is a mass of protoplasm.

The cell, already a complex formation, has as its starting point a solid mass of protoplasm. This first and transitory state

*Virchow, *La Pathologie Cellulaire.* 4th edition, Paris, 1874.

soon gives rise to more complex states. The first degree of complication is the formation of the *nucleus,* by the condensation of protoplasmic particles, a sort of a nebula which becomes more and more clearly defined. Then the protoplasm is clothed in a denser layer, the beginning of the membranous *envelope* that will later become distinct. Here is a second stage, a second degree of complexity. The cell then appears before us as a solid little body with a nucleus and a cortical layer.

Development can then stop here; the transitory form can become a permanent form, in animals as well as in plants. Such are the bodies Haeckel has called *cytodes,* two forms of which exist:

1. The *Gymnocytode,* a mass of albuminoid material without appreciable structure, without definite form, devoid of all organization, showing no signs of any differentiation into parts. This mass is finely granular; the granulations are observed all the way to the periphery.

2. The *Lepocytode* is a somewhat more complicated form, already presenting a primary degree of differentiation. There is a cortical layer, or envelope; the peripheral protoplasm is distinguished from the central; this latter, for example, is granular and more fluid, while the cortical protoplasm is without granulations, brilliant, refringent, homogenous, and resistant, fulfilling the function of an envelope.

The cytodes, as we shall see later (see Lecture 8) can form individual and complete living beings. Haeckel thus called them *monera.* Within recent years the study of these rudimentary beings has taken on great importance and great development at the hands of Haeckel, Huxley, and Cienkowsky. *Protogenes primordialis,* discovered in 1864 by Haeckel, the *Bathybius haeckelii,* discovered in 1868 by Huxley, are gymnocytodes. *Protomyxa autentiaca* and *Vampyrella,* studied by Cienkowski in 1865, are Lepocytodes.

Bathybius haeckelii has been found at depths of 4000 to 8000 meters in the fine chalky silt of the ocean. It was described as a sort of mucilagineous mass formed of lumps, some rounded, some amorphous, forming at times a viscous network covering fragments of stone or other objects (See Figs., Lecture VIII.)

Such a mass of protoplasm, granular and without a nucleus, is thus characterized only by itself, by its own constitution; it has no definite and characteristic form. It is nevertheless a living being;

its contractility, its property of nutrition, of reproducing by segmentation, are proof of this.

These observations, once disputed, particularly as concerns Bathybius, have received complete confirmation in recent work carried out in these last three years.

The reproduction of these beings by scissiparity has been observed in Protamoeba and Protogenes when these mucous bodies have attained a certain size (See Figs. Lecture VIII). The mass that constitutes them forms a constriction and divides into two halves, each of which becomes rounded and behaves like a distinct being; it has been possible to say that "here reproduction is only an overgrowth of the organism, that exceeds its normal volume."

Segmentation sometimes takes place into four parts *(Vampyrella)* or into a greater number, but the process of reproduction is always equally simple.

Among these protista there is such an intimate mixture of animal and plant characteristics that they cannot be clearly assigned to the one more than to the other, and certain naturalists have formed a third kingdom out of them intermediate between the animal kingdom and the plant kingdom (Haeckel, p. 369).

But these bodies can equally well represent transitional states of organisms passing on to a higher level. Starting from this gymnocytode state, certain organisms become leptocytodes, later acquiring a nucleus, become true cells, at first naked, but later provided with envelopes and in a word, complete.

In a still more advanced state, the protoplasm, after having formed its tegument, and its nucleus, excavates vacuoles filled with *cellular fluid*. This is what happens in plants. Then these vacuoles coalesce into a central lake, so that the protoplasm finds itself more or less uniformly pushed out, along with its nucleus, to the periphery. Thus it forms a layer lining the envelope internally. Hugo Mohl was the first to see this subtegumental layer; he understood the importance of its role and gave it the name of *primordial utricle*. The phytoblast thus takes on the form of a hollow sac and well merits the name of cell.

It is in this state that cells were first seen. The English botanist Grew (1682) called them *vesicles*. Malpighi (1686), *utricles,* the French botanist de Mirbel (1808) first employed the name *cells* to characterize them. It was not until 1831 that the celebrated English botanist R. Brown considered the nucleus *(kernel, spherid* of de Mirbel) as an essential part of the cell; Schleiden (1838) called attention to the existence of the *nucleolus;* from then on all parts of the cell were known.

Finally, and it is the last stage in this development, the protoplasmic layer gets thinner and thinner and finally disappears. The cell is then dead; it is a cadaver. Hugo Mohl (1846) clearly perceived this essential difference between cells that have a primordial utricle and those that do not. "Only the first are in the stage of growth, of producing new chemical combinations, of forming, in favorable circumstances, new cells. The others are henceforth incapable of all further development; they no longer serve the plant except by their solidity, by their power of imbibition of water and by their particular form." The protoplasm is in fact the living body of the cells; it forms all the other parts and all the substances that the plant contains. The nucleus and the envelope are improvements produced by the protoplasm, the only living and working material.

The preceding considerations establish therefore that life in its simplest stage, stripped of its complicating accessories, is *not bound to a fixed form,* for the cytodes has none, but to a *definite physicochemical arrangement or composition,* because the material of the cytode is a mixture of albuminoid substances possessing quite constant characteristics. Here therefore the morphological concept disappears before the concept of the physicochemical constitution of living matter.

This matter is the *protoplasm*. E. van Benedum proposed to call it *"plasson"* and Beale, *"bioplasm."* It can be said with Huxley* that it is the *physical basis of life*.

The ultimate degree of simplicity that an isolated organism can exhibit is therefore that of a granular mass without dominant form. It is a body, not morphologically defined as it was thought that all living beings ought to be, but chemically, or at least because of its physicochemical constitution.

*Huxley, *Les Sciences Naturelles et les Problèmes qu'elles font Surgir*. Paris, 1877.

It is not just a few exceptional beings that present themselves in a form so simplified; all beings, all higher organisms, appear to be temporarily in the same situation. The egg, in fact, finds itself at a certain time in the same situation, when it has lost the germinative vesicle, before being fertilized.

The cell, the anatomical element found as the basis of all organization, animal or plant, is nothing else but the primary definitive form of life, a sort of a mould in which the living matter, the protoplasm, is enclosed. Far from being the ultimate degree of simplicity that can be imagined, the cell is already a complicated apparatus. This body possesses an *envelope,* a *cellular membrane* or cortex, granular contents, the *protoplasm* or *cell body,* a circumscribed mass included in the protoplasm, the *nucleus* or kernel, which itself contains little corpuscles or *nucleoli.* The designation of cell is inexact; it is applied in fact to a body that undergoes a series of successive and continuing transformations; it is in one of its transitional states (the only one that was known at first) that it presents the form of a sac, implied by the name cell. Today the name *phytoblast* is substituted for that of plant cell. At the start, and in its greatest degree of simplicity, the phytoblast appears to us as a small rounded mass of a substance that is more or less finely grained, without a condensed nucleus or distinct wall. This substance, call *sarcode* by Dujardin, who had the animals particularly in mind, is commonly designated by the name *protoplasm.* In the beginning a phytoblast is thus a naked, spherical lump of protoplasm; the animal cell originally shows the same structure *(gymnocytode* of Haeckel) .

In its most rudimentary state, life resides in this mass of protoplasmic substance. This state, which is the simplest and the earliest in which the element is found, does not ordinarily persist. As we have said, it is a starting point which will become complicated by successive differentiations.

III. PLASTIDULAR THEORY

We have just seen how we have been led step by step to localize life within a substance defined by its composition and not by its form, the *protoplasm.* Let us look at the ideas that have been held about this substance; then we shall examine the problem of its creation or its formative synthesis.

What is the physical nature of protoplasm? It was first thought that this substance was homogenous, without appreciable structure.

In 1870 a change took place in the ideas, and the *plastidular* theory came into being. A final step was taken during the last two years by the research of a number of micrographers, Bütschli, Strassburger, Heitzmann, and Frohmann.

Naked protoplasm is by no means the final limit attainable by microscopic analysis. In many cases protoplasm exhibits a sort of framework, formed from a network of fine granulations joined by very delicate filaments; these are the *plastidules.* The plastidular theory is thus the ultimate point to which histology can carry the concept of living beings. When Heitzmann and Frohmann examined the basic tissue of cartilage or the nuclei of blood corpuscles in crayfish, they observed very clear fibrils, arranged in a plastidular network, at the intersection of which small granular masses were found. (See Figs., Lecture VIII) .

Haeckel accepted the existence of these plastidules as a general fact. He regarded them as the ultimate elementary constituents of the monera, the irreducible bodies to which analysis could lead. This element would be active, enjoying vibration and undulatory movements, the plastidular movements. Haeckel attributed to them the physical properties of the material molecules, and in addition one vital property, *memory,* or the faculty of maintaining the kind of motion by which their activity was manifested. This notion of the faculty of remembrance, or memory, considered as the elementary property of organic particles had already been put forward in the last century by Maupertius, in his Venus Physique, and was recently defended by the physiologist Ewald. Finally, an American physician, Ellsberg, tried (1874) to rejuvenate Buffon's theory of reproduction by substituting the plastidules, which have a more certain existence for the organic molecules invented by the great naturalist.

It is obviously necessary to wait for more ample confirmation to establish the general applicability of the facts described above concerning the *complexity of the structure of the protoplasm.* It can nevertheless be said that at the present time a whole series of studies argues in favor of this complexity: such are the studies of Strassburger on the nuclei of plant cells during cell division, those of Bütschli on the nuclei of blood corpuscles, of Weitzel on the cells of the inflamed conjuctiva and the cells of the frog's skin, of Balbiani on the epithelial cells of the ovaries of certain insects such as Sthenobothrus, of Hertwig on the hen's egg, and of Fol on certain invertebrate eggs.

Later, when we take up the general morphology of the living beings and the genesis of their tissues (see Lecture VIII), we shall enter into the details of these studies. For the moment we shall mention only the principal observation, which we owe to Strassburger. This author observed the nuclei of the ova of certain Abietineae at the moment when the cells begin to divide to form the embryo. The nucleus is elongated; masses of material connected by filaments form at the two extremities. In the middle of these filaments there appear granulations whose whole forms a disc (nuclear disc); soon the granules divide in two groups, and each half migrates toward the corresponding pole, where it increases the size of the polar mass.

A granule now appears, de novo, in the center of the filament; the whole forms a *cellular plaque* or disc which soon divides into two parts which go to rejoin the polar masses.

Here is a phenomenon that reveals to us a very complex constitution of the nucleus.

Now this is not an isolated observation at all. Identical facts can be established in the algae, the spirogyra, and from now on it must be accepted that they have a true general applicability within the plant kingdom.

The animal kingdom affords similar examples. Here we confirm once again a constant parallelism between plants and animals, by virtue of which all the essential phenomena are found to be identical in the two kingdoms. Bütschli, studying the division of blood corpuscles in the embryo, encountered the fibrillary tracts, the nuclear plaque, that divides in two, and the cellular plaque, whose segmentation brings on that of the nucleus. Balbiani has observed the same in Sthenobothrus, and he considers the equatorial granules to be nucleoli (See Figs., Lecture VIII).

These observations and the generalization to which they are susceptible, have as a consequence to make a complex body of the nucleus, a mass of protoplasm hitherto considered as simple, both from the anatomical point of view and the physiological point of view.

On considering the cell, which is a rudimentary living being, the two types of essential phenomena, of organic creation and vital destruction, ought to be found therein. Now the preceding studies, the microscopic studies of the *nucleus,* and our own observations, seem to localize the one and the other order of phenomena in a different part; in the protoplasm on the one hand, and in the nucleus on the other hand.

The protoplasm is the agent of the manifestations of the cell; vital manifestations which become apparent in the activities of the tissue where they assemble and multiply. The functional phenomena or vital expenditures *should therefore have their seat within the cellular protoplasm.*

The nucleus is the apparatus of *organic* synthesis, the *instrument of production,* the *germ of the cell.* We have observed (see Lecure VI) that the formation of animal starch is associated with the existence of the nucleus in the glycogenic cells in the amnion of ruminants. The ideas acquired by the most competent histologists lead to this interpretation. One knows of the function that devolves on the nucleus in cell division, and the initiative that attaches to it.

Numerous observations confirm this concept that makes the nucleus the cellular reproductive apparatus. Ranvier has established in the lymphatic corpuscles of the axolotl the presence of a true budding of the nucleus, which, at first rounded, sprouts prolongations at different points around which the protoplasmic substance groups in such a way that each of these prolongations appears soon as the beginning of a new organism, and as the first stage of a lymphatic corpuscle of the second generation.

R. Hertwig has established the same phenomenon of budding of the nucleus in one of the acinetae, *Podophyrya gemmipara,* in which nuclear growth is the starting point and the signal for the multiplication of the animal. The cells of the Malpighian vessels in insects exhibit the same phenomena. It is not necessary to multiply examples to perceive their general applicability.

The penetrating studies that some histologists have recently carried out on the constitution of the nucleus of the cell have revealed the complexity of this element, incorrectly considered to be simple. N. Auerbach distinguishes four parts within the nucleus:

The envelope
The nuclear fluid
The nucleoli
The granules

Of these elements the one of greatest importance is the *nucleolus*. The nucleolus is a formed corpuscle that R. Brown pointed out in 1831 in plant cells. Two opinions exist as to the nature of the nucleolus. The one consists in considering the nucleolus as a solid protoplasmic mass, the true germ of the cell. Auerbach, Hoffmeister, and Strassburger accept this viewpoint.

The other opinion consists in regarding the nucleolus as a hollow lacunar mass of *vacuoles, nuclear vesicles,* or *nucleolules.* Balbiani, who attracted the interest of histologists to this structure, has deduced from this a physiological interpretation of the role of the nucleolus. He looks upon it as an organ of *nutrition,* as a *sort of heart.* In the nucleoli of a great number of cells Balbiani has discovered movements that can be classified in two types: (1) amoeboid movement analogous to those of protoplasm, (2) contractile movements of the vesicles or vacuoles located in the homogeneous mass of the nucleolus.

The amoeboid movements of the nucleolus have been seen by Balbiani in the germinative spot (representing the nucleolus) in the eggs of certain arachnids, in particular, *Epeire diadema.*

This observation has been confirmed by those of a great number of histologists; by Lavalette Saint-Georges in a larva of Libellule, by Auerbach and Eimer in fish, by Al Braun in *Blatta orientalis.* Mecznikow has encountered these same movements in the cells of the salivary glands of ants, and finally, W. Kühne has noticed them incidentally in the corpuscles in the pancreatic juice of the rabbit.

The second kind of nucleolar movement consists in the contraction of the vesicles. They are most evident in the ovule of the common daddy-longlegs, *Phalangium,* and a myriapod, *Geophilus longicornis.*

The nucleolus is almost as constant an element as the nucleus. The absence of the nucleolus, the *anucleolar state* of Auerbach, is most always transitory and temporary; this is what happens during segmentation of the egg. Some elements have only a single nucleolus; nerve cells, and the cells of the spinal cord are of this kind. In mammals and birds there is always a number of nucleoli in the nucleus, varying from 4 to 16. In fish this number is singularly increased; in the germinative vesicle of these animals a number of nucleoli varying from 150 to 200 for each nucleus is found.

CONCLUSION

In the rapid review of all the studies that have appeared recently on these delicate subjects, we have seen the different forms in which the essential material of the organism, the protoplasm, can present itself. After being considered as a material with a very simple constitution, it is now regarded as a very complex structure. All the problems of organic origin, and all the questions associated with it, are by no means resolved. Nevertheless we can stop with this general conclusion, that the materials of the living edifice represent different forms of a unique substance, the depository of life, identical in animals and plants. It is in the protoplasm, the sole active and functional material, that we must seek the explanation of life, not only of the chemical phenomena of nutrition but also of the higher vital reactions of sensibility and movement.

SIXTH LECTURE

Chemical Theories – Syntheses

Colorless Protoplasm and Green, or Chlorophyllian, Protoplasm

SUMMARY: Protoplasm and organic creation—Generalities—Chemicophysiological synthesis—Elementary constitution of organized bodies—Creative synthesis is necessarily chemical, but it has special processes—Green, or chlorophyllian, protoplasm and colorless protoplasm—They cannot serve to delimit the animal kingdom from the plant kingdom.

I. Role of Chlorophyllian Protoplasm in Organic Synthesis—It produces the synthesis of ternary bodies under the influence of light—The experiment of Priestley is the starting point of this theory—Hypotheses of the chemists on the subject of syntheses in green protoplasm—Green protoplasm derives its energy from solar radiation.

II. Role of Colorless Protoplasm in Organic Syntheses—It produces complex syntheses—Pasteur's experiments—It cannot incorporate carbon directly, however—Colorless protoplasm uses caloric energy—Status of the question of organic syntheses; new hypotheses—Hypothesis of cyanogen—Chemical synthesis and vital force.

III. Synthesis in Particular—The best known example is amylaceous or glycogenic synthesis—Discovery of animal glycogenesis—Phenomena of synthesis of starch and destruction of starch—Principal characteristics of glycogenic synthesis in animals and plants.

WE HAVE SEEN ABOVE that the essence of life must be separated from the form of its substratum; it can be manifested in a material that has no definite morphological characteristic. It is in this material, the *protoplasm,* that the vital activity resides, independently of the morphological state that it displays, and the moulds in which it was fashioned. Protoplasm alone lives or vegetates, works, makes products, breaks itself down and regenerates itself without cease, and is active by virtue of its substance and not because of its form or shape.

The fundamental phenomenon of *organic creation* consists in the formation of this substance, in the *chemical synthesis* by which it is constituted from materials from the external world. As to the *morphological* synthesis that fashions this protoplasm, it is, so to speak, an epiphenomenon, a consecutive event, a stage in that infinite series of differentiations that lead up to the most complex forms; in a word, an elaboration of the essential phenomenon.

Thus, Lavoisier was right, when while proclaiming the difficulty of the problem of organizational creation, and recognizing that it was surrounded by an impenetrable mystery, he nevertheless claimed it as a chemical phenomenon, a phenomenon that chemists from then on ought to undertake to study. He proposed for the Académie des Sciences to encourage and initiate studies by the establishment of prizes awarded to authors who made advances in this direction.*

The problem of organic creation or vital synthesis would thus have as its first stage, and as its essential condition, the chemical synthesis of protoplasm.

The chemical constitution of protoplasm cannot at present be determined; the formula $C_{18}H_9N_2$ by which it has been represented is completely illusory. Protoplasm is a complex mixture of immediate principles, albuminoid materials and others, poorly understood containing as primary elements carbon, hydrogen, nitrogen and oxygen, and as secondary elements certain other simple substances. In a word, there are to be found, just as in the *blastema,* quaternary and ternary bodies and minerals.

---

*See the note of Dumas, *Leçons de la Société Chimique.* 1861, p. 294.

The elements that chemistry has revealed to us as entering into the constitution of the most complex organisms are but few. There is no special substance or elementary living particle, such as Buffon envisioned to explain the difference between living beings and inanimate bodies. The only elements that enter into the material composition of the higher beings, of man, for example, are fourteen in number. They are:

Oxygen	Chlorine
Hydrogen	Sodium
Nitrogen	Potassium
Carbon	Calcium
Sulfur	Magnesium
Phosphorus	Silicon
Fluorine	Iron

These are the elements that chemical synthesis sets in motion, and which through successive combinations, come to form the substratum of life.

These elements combine with each other, indeed, to form binary, ternary, quartenary, and quinary combinations, etc.; these assemble to constitute the original living substance, *blastema, plasma,* or *protoplasm,* in which all the essential acts of life are manifested. At a higher level, materials assume a morphological character, and constitute the anatomical element, the cell, and still further along, complex organisms.

The problem of the mechanism of these organizational syntheses is very far from solution; it has not yet even been stated properly; here we shall try no more than to define the question and make known the state of the science on this subject.

We have said that Lavoisier was right in delegating to chemistry the explanation of the phenomena of the organization of living beings. From the time that he expressed himself so clearly, synthetic chemistry has in fact made considerable progress. Plant essences, fats, and alcohols have been totally synthesized. The great work of Berthelot on synthesis has disclosed the possibility of great advances in this direction; the recent investigations of Schützenberger make it probable that it might even be possible to reconstitute albuminoid substances artificially, which are rightly considered as the highest level of vital synthesis.

But these very advances in chemical synthesis oblige us to ask ourselves if physiology can expect from them the solution of the problems of physiological synthesis. In other terms, it is a matter of knowing whether the processes by which the chemists have formed these natural compounds are the exact duplicate of those employed by a nature; whether the chemical synthesis which forms the organic substances in the body is similar to that of our laboratories.

It seems to be otherwise. The physiological or natural processes, although they are included among the laws of general chemistry, do not necessarily resemble those that the chemist sets at work; they are generally different; they are special. What is already known about the transformations and the syntheses of fats, sugars, and starches gives credence to this point of view, which I have maintained so long. This is also the opinion of chemists who are best acquainted with synthetic methods, and who have carried out the most remarkable work in this field.

Everyone knows for example that Chevreul was the first to carry out the analysis of the fats. He showed that they are formed by the union of glycerine with one or more fatty acids. Starting with these products Berthelot reconstituted the fatty materials and effected their synthesis. Now neither Chevreul nor Berthelot draw from their work the conclusion that fats are constituted in the living being, by the same processes. In a word, they do not think that fat is formed in animals or plants by the necessary union of preexisting fatty acids and glycerine.

More recently Schützenberger has studied the composition of albuminoid materials. He seems to have succeeded in carrying out their immediate analysis, or rather, *an* immediate analysis. By treating the albuminoid material with a solution of baryta at 150 degrees, he obtained definite crystallizable principles. These substances obtained by decomposition can be arranged in three series:

1. Ammonia, carbonic acid, oxalic acid, and acetic acid; these substances in a constant proportion for each given albuminoid substance

2. Secondly, crystallizable nitrogenous compounds belonging to two series which have leucine and leuceine as their prototypes

$$C_nH_{2n+1}NO_2 \quad (n = 3, 4, 5, 6, 7)$$
$$C_nH_{2n-1}NO_2 \quad (n = 4, 5, 6)$$

3. Compounds such as pyrrole, tyrosine, tyroleucin, and glutamic acid.

The differences between the various albuminoid materials appear to relate first to the relative proportions of these three kinds of substances and then to the nature and the relative proportion of the substances belonging to the second group.

The analysis having been made quantitatively, that is to say, weight for weight, Schützenberger thought that it would now be possible to represent the constitution of albumin in a chemical formula:

$$6C_9H_{18}N_2O_4 = \underbrace{C_6H_{13}NO_2}_{\text{Leucine}} + \underbrace{C_6H_{11}NO_{12}}_{\text{Leuceine}} + \underbrace{C_5H_{14}NO_2}_{\text{Butalanine}}$$
$$+ C_5H_9NO_2 + \underbrace{4(C_4H_9NO_2 + C_4H_7NO_2)}_{\text{Amido-butyric acid}} + \text{Water}$$

For each nitrogen-containing substance there would be a similar formula.

Is this to say that in the very opinion of the author of these laborious and remarkable studies, the synthesis of albumin takes place in the organism by successive combinations of these elements? By no means. Nature seems to proceed along quite different lines.

It is, indeed, always chemical combinations that are made and unmade, but the organism has special processes, and only the study of the living being can enlighten us about the mechanism of the phenomena for which it is the theatre and about the special agents it employs.

Here we ought to point out something important. We do not witness the direct synthesis of the primitive protoplasm, any more than any other primary synthesis within the living organism. We observe only the development, the growth of the living material but it has always been necessary for some sort of vital leaven to be the starting point. At the beginning of the development of any living being whatsoever, there is a preexisting protoplasm that comes from the parents and has its seat in the egg. This protoplasm grows, multiplies, and engenders all the protoplasms of the organism. In a word, just as the life of the new being is only the continuation of the lives of the beings that preceded it, in the same way its protoplasm is only the extension of the protoplasm of its ancestors. It is always the same protoplasm, it is always the same being.

Protoplasm has the property of growing by chemical synthesis, it renews itself after organic destruction. These two properties constitute the life of the protoplasm, which we have to examine.

Some physiologists have seemed to believe that two kinds of protoplasm must be distinguished, behaving differently; the *colorless protoplasm* of animals and the *green protoplasm* of plants.

In reality, even in respect of color, animal protoplasm cannot be distinguished from plant protoplasm. The protoplasm of plants, like that of animals, is capable of impregnating itself with green material, or chlorophyll, in certain circumstances. This material, so important in its functions, can appear or disappear within the preexistent protoplasm, according to external circumstances. If, for example, certain portions of a green leaf are covered by an opaque screen, the parts thus withdrawn from the action of light lose their color; the chlorophyll disappears, the protoplasm survives by itself.

Instead of saying, therefore, that two varieties of protoplasm exist, it would be more exact to say that protoplasm, according to circumstances, is laden with green material or contains none at all; above all there is no need to consider plant protoplasm as opposite to animal protoplasm. To us this would be most inaccurate; in fact, a third at least of the plant species are devoid of chlorophyll; in a given plant all parts removed from the action of light are in the same circumstance; finally, as we shall see later on, the lower animals, *Euglena viridis, Stentor polymorphus,* etc. (see Plate, Figs. 1 and 2) possess this same substance.

At any rate, while reserving the question of the original unity of the protoplasm, and on condition of relegating the chlorophyll with which it is mixed to the status of a *product,* it is practically permissible to distinguish green protoplasm from colorless protoplasm.

In fact, in certain cases, these two protoplasms seem to behave in an entirely different way from the standpoint of chemical syntheses.

I. GREEN, OR CHLOROPHYLLIAN, PROTOPLASM

Chlorophyll is found in the greater number of plants, in the parts exposed to light. It appears disseminated within the cellular protoplasm in the state of granules with an average dimension of 0.01 mm; sometimes, however, it seems to be in true solution.

The botanists agree that this substance is a product of the activity of protoplasm; for in germinating seeds or in etiolated plants brought back into the light, this material is seen to reappear within the protoplasm which had never ceased functioning. Studying the phenomenon more closely it was believed to be possible to say that the chlorophyll is generated in the layer of protoplasm surrounding the nucleus of the cell and its appearance was related to the influence of the nuclear protoplasm.

The facts relative to *animal chlorophyll* are no less interesting, although they might be less-known. In 1844 Morren had begun to study the respiration of several green organisms which obviously did not belong to the plant kingdom. But it was especially F. Cohn in 1851, Stein in 1854, and Balbiani in 1873 who gave a more solid basis for our knowledge in this regard.

F. Cohn established the presence of grains of chlorophyll in an infusoria, *Paramecium bursaria;* these grains are located in the internal, more fluid part of the cortical layer (the body wall). This fluid layer is in a constant rotating movement, in which the green granules participate. These granules show reactions similar to those of plant chlorophyll. Concentrated sulphuric acid first gives them a greenish-blue coloration which gradually becomes more intense and finally changes into blue with dissolution of the granules.

Stein has confirmed these facts; he has given greater definition to the location of the chlorophyll granules in the protoplasm that forms the general mass of the body, apart from the digestive tube and the cortical region. He has, moreover, seen species that were at times colorless, and at other times colored green, such as *Spirostomum ambiguum, Ophrydium versatile, Epistylis plicatilis, Stentor polymorphus,* etc. In many of the flagellate infusoria, *Euglena viridis, Cryptomonas, Chlamydococcus pluvialis,* and *Trachelomonas,* the green material shows up in the amorphous state or in the state of very fine granules. In these infusoria, as in plants, the chlorophyll is transformed at certain periods, especially during encystment, into a yellow-red pigment; it reverts to green when humidification restores the animal to active life.

In 1873 Balbiani (see Plate I, Figs. 1 and 2) observed in *Stentor polymorphus* (the green variety) the multiplication of granules of chlorophyll within the interior of the animal's body, by division into two and three, as it takes place in plant chlorophyll. Outside of the infusoria listed above, green corpuscles are found in the substance of the body in various other species of animals; *Hydra verte,* a turbellarian worm, *Vortex viridis,* and a gephyrean, *Bonnellia viridis.*

These facts demonstrate how little basis there would be for attributing green protoplasm exclusively to plants, while colorless protoplasm would characterize the animal.

What is the role of green protoplasm in organic synthesis?

It is green protoplasm which, according to the ideas in favor at present, carries out the synthesis of *ternary hydrocarbon compounds.* It would be the sole agent of the synthetic combinations of carbon, the sole means for the introduction of this substance into the plant and animal organism.

The celebrated experiments of Priestley have been the starting point of our knowledge in this regard. Ingen-Housz, Sennebier and Th. de Saussure have determined the conditions for this experiment, and have revealed the synthetic action exerted by the green material. Following their work it has been accepted that chlorophyll has the faculty of reducing carbonic acid under the influence of the solar rays and giving rise to a liberation of oxygen. At the same time the carbon is combined with different elements and constitutes carbohydrate or combustible materials which are stored in the green organs.

How does this action take place? In this regard there are nothing but more or less plausible suppositions. One tended to think:

> . . . the normal hydrate of carbonic acid is split into oxygen and methyl aldehyde under the action of chlorophyll; the aldehyde sextuples to give sugar, which in its turn, by duplication or triplication and the loss of water, gives cellulose: oxydation of these bodies furnishes the fats and acids; the action of ammonia derived from the reduction of nitrates, forms from the above radicals the various plant alkaloids and the albuminoid materials.

For these hypotheses, which he first reviews, Gautier *(Revue Scientifique.* 10 February, 1875) has substituted others which seem to be more in accord with the small number of known facts.

It must first of all be admitted that the green material chlorophyll is not intimately incorporated and firmly combined with the protoplasm itself; it is simply disseminated throughout the protoplasmic mass, from which it can be extracted by a large number of neutral solvents.

This green protoplasm is the agent of a multitude of carbon syntheses whose products, made during the day under the action of the sun's rays, are utilized as building materials by all the colorless parts of the plant.

According to Gautier, two states of chlorophyll should be distinguished:

Green chlorophyll
White chlorophyll

In the etiolated parts that become green again in light, the substance that can give rise to chlorophyll is present, for it is sufficient to treat them with sulphuric acid to see them instantly become green. Gautier believes that under the influence of the oxygen of the air white chlorophyll passes into the state of green chlorophyll, and inversely, that *green chlorophyll* passes into the state of *white chlorophyll* under the influence of nascent hydrogen; the experiment can be made and repeated easily.

The two substances, green chlorophyll and white chlorophyll, would have the relationship to each other of indigo blue and indigo white. White chlorophyll would have the remarkable aptitude of reducing oxygenated bodies, to combine their oxygen with its hydrogen. Moreover, green chlorophyll would have the property of decomposing water under the influence of the sun's rays, as it has the property of decomposing carbonic acid. It would become white chlorophyll by taking in hydrogen and setting oxygen free. The white chlorophyll would yield its hydrogen to the carbonic acid; it would thus operate in the synthesis of carbon compounds and return to the state of green chlorophyll.

Thus by a constant alternating movement, the chlorophyll would assume the green and the colorless state, decomposing water and giving off oxygen when it passes from the green to the colorless state, and carrying out the synthesis of carbon containing products on returning from the colorless state to the green state.

This is the first part of the hypothesis. It is still far from being verified or based on facts, but it is not contrary to any that are known.

This is the second: what are the basic materials on which the green or white chlorophyll exert their actions? It is a mixture of carbonic acid and water, $nCO_2 + mH_2O$. From the reduction of this mixture, thanks to the hydrogen of the chlorophyll, there are derived alcohol, glycerine, ordinary aldehyde, glycollic and glyoxalic acids, glyoxal, and oxalic acid. In a word, all the "organic ternary bodies can be formed by this simple mechanism of deoxidation by the chlorophyll granule, more or less extensively according to the influence of the light rays, and the various associations of water and carbonic acid which the protoplasm permits to penetrate into the organ of reduction."

Glucose would be the first of these principles to be formed and the basic material for nearly all the rest. Through combination with carbonic acid and the loss of water, glucose could give pyrogallic acid and gallic acid, which in the new growth of springtime is in fact closely associated with glucose: in a word, a series of acids which can, inversely, return to the state of a sugar under the influence of the life of the colorless cells.

Thus in the colorless parts phenomena take place which are exactly the opposite of those that occur in the green parts. It is in fact a general tendency among chemists to accept this reversal, similar in its mechanism, although opposite in its direction, from existing plant materials to the immediate principles from which other cells had derived them.

Here are some of the ideas that the chemistry of our times has set forth on the role of green protoplasm in the synthesis of immediate products.

These concepts are strongly impregnated with what can be called *synthetic chemistry. Natural chemistry* is perhaps entirely different; it would be possible for example that all the syntheses invented by the chemists are without reality, and that the immediate principles are produced by way of decomposition or cleavage of a single and identical material, the protoplasm.

However this might be, and to stay within the facts, it can be said the green protoplasm seems incontestably to form organic carbon compounds.

Under the influence of what force and by what energy are these phenomena carried out? Where does the cell with green protoplasm derive the chemical energy necessary for the decomposition of carbonic gas?

It is agreed that it is from the solar radiation. The sun is the prime mover of all these phenomena, the source of the kinetic energy they utilize.

II. COLORLESS PROTOPLASM

We have just seen that protoplasm is capable in certain circumstances of loading itself with a green material, chlorophyll. But protoplasm can remain without color in a great number of plant elements. Colorless protoplasm is, even less than green protoplasm, an exclusive attribute of one of the kingdoms. Animals and plants possess it as an essential, primordial element, creator and generator of all the others.

What is the role of this protoplasm? It could be to produce all the substances that exist in animals and plants, though with other elements as starting points, and with a kinetic force as agent other than that of green protoplasm.

Pasteur's experiment on this subject is fundamental. It shows that colorless protoplasm can fabricate, without the help of chlorophyll or of solar radiation, the most complex immediate principles; protein materials, albumin, fibrin, cellulose, fatty materials, etc.

Pasteur *(Comptes Rendus.* 10 April 1876.) makes up a culture medium from the following principles:

Pure alcohol or acetic acid
Ammonia (a pure crystallizable salt)
Phosphoric acid
Potassium
Magnesia
Pure water
Gaseous oxygen

Here there is not a single substance that is not taken from the mineral kingdom, for even the most complex, alcohol, can be produced wholly from elements taken from the mineral kingdom, as Berthelot has shown.

Into this medium of such a simple constitution, without albumin, without organized products, is placed a particle of *Mycoderma aceti,* of so to say no weight, of an insignificant mass.

In the absence of all green material, in the dark, the particle of mycoderma produces in this medium a considerable quantity of new cells of *Mycoderma aceti,* of as great a weight as could be desired. In this yield the most varied and complex materials of the organism are encountered. The most varied and complex materials of the organism:

Protein materials
Cellulose
Fatty materials
Coloring materials
Succinic acid, etc.

The living cell has therefore no need of chlorophyll or green material, nor of solar radiations in order to fabricate these most complicated immediate principles of the organization.

Pasteur has furnished a second example, by cultivating vibrios, that is to say, still higher beings, in the dark, without green material and moreover, without gaseous oxygen. The culture medium was constituted as follows:

Lactic acid
Phosphoric acid (as a pure crystallizable salt)

Ammonia
Potassium
Magnesia

A few vibrios are sown in this medium, in such a small amount that it cannot be measured.

These beings develop with prodigious activity, and as great a weight of them as desired can be obtained, containing:

Cellulose-like materials
Protein materials
Coloring substances
Alcohols
Butyric acid
Metaacetic acid, etc.

It can in consequence be said that colorless protoplasm has accomplished some very advanced syntheses.

Nevertheless, between these syntheses accomplished by colorless protoplasm and those that green protoplasm produces, there are two differences. To begin with, in the first case, a sufficiently complex carbonaceous principle, such as alcohol, acetic acid, or lactic acid, is necessarily provided as a starting point; life would not be possible if carbon were given a more simple state, for example in the state of carbonic acid. Chlorophyll alone can complete the synthesis of carbonaceous or ternary principles, starting from simpler or more saturated compounds such as CO_2. Colorless protoplasm, from this starting point, can perform the most complicated quaternary syntheses.

Another difference results from the energy employed.

Green protoplasm makes use of the energy of the light rays, that is to say of the kinetic energy of the sun.

Colorless protoplasm makes use of the caloric energy that has its source in the carbonaceous foodstuff; this must fulfill only one criterion; it must not be saturated with oxygen and must in consequence be able to provide heat by saturation or oxidation.

Pasteur would comprehend in the last analysis, and as a hypothesis, that colorless protoplasm might under the influence of electric waves or some other form of kinetic energy, decompose carbonic acid and assimilate carbon to form it into ternary synthetic products.

However this might be, in the present state of things, different roles are attributed to the two protoplasms; the green prepares ternary carbonaceous compounds, the colorless, from this starting point, makes quaternary nitrogenous principles. In a plant, the green cells would thus work for the colorless ones.

If a plant has no green parts, it can live only on condition of finding, fully prepared in the external environment, the principles that were elaborated beforehand by the chlorophyll of some other plant. This would appear to be the case with parasitic plants such as molds, slimes, and monocellular beings, which must find on the organism that carries them or in the medium that bathes them, these same indispensable principles, the source of their protoplasmic activity.

It is in this sense that Boussingault, and with him several chemists, have been able to say that plants (it should be said, the green material) alone are capable of providing living beings with carbon, and in consequence of creating the immediate principles with the help of inert, mineral elements, derived from air, water, and earth. This creative power would be possessed by chlorophyll alone, under the influence of the sun. "If solar radiation were to stop, nor only the plants with chlorophyll but even the plants that are void of it would disappear from the surface of the globe."

The experiment of Pasteur who takes mineral products and a *laboratory product,* alcohol, as a culture medium, redresses what is perhaps excessive about this view. *Mycoderma aceti* and the vibrios that developed in the artificial medium made up by Pasteur, had no need for any antecedent chlorophyll-containing plant, nor for solar radiation.

All the explanations that we have given relating to the processes of organic synthesis indicate the general way in which present day thinking conceives of these phenomena. But their exact mechanism, we have already said, could be quite other than these hypotheses imagine it to be. Here as in many other cases, chemical explanations tell us how things could be, more than they show how things really are. Only experimentation carried out on the living being can tell us.

From the physiological standpoint, one would be justified in thinking that in the organism there is only one synthesis, that of the protoplasm which grows and develops from the appropriate materials. From this complex body, the most complex of all organized bodies, there would be derived by subsequent cleavage, all the ternary and quaternary compounds whose appearance we attribute to a direct synthesis.

This concept, which would have all the products of the organism derived from a single compound, protoplasm, is itself still a hypothesis. It would nevertheless not be difficult to collect a certain number of facts that are in accord with it. An argument in its favor would be for example, the maintenance of the fixed constitution of the organism, with a varied diet. The products of the organism do not change sensibly under the influence of the diet, and this would be explained perfectly if the materials were derived exclusively from a protoplasm, itself always identical.

Finally, we can do no more than mention a final hypothesis on the origin of living material, although it has been the object of considerable development on the part of its author.

Relative to organic creation Pflüger *(Archiv für Physiologie.* vol. X, 1875.) proposed a hypothesis that could be called the *cyanic* hypothesis. According to Pflüger it is not carbonic acid, water vapor, or ammonia that would preside over primitive organic synthesis at the start of life. "These materials," he says, "are the results and the termination of life, rather than its beginning, which is in accord with their great stability." The origin of living material, according to the author, ought to be sought in cyanogen.

But first what might be the origin of this cyanogen? It would be the oxygenated compounds of nitrogen, which in certain climatic conditions, storms, etc., can form cyanic combinations. Pflüger explains that, when the world was incandescent, cyanogen could have been formed, and he always points to fire as the force that has produced by synthesis the constituents of the molecule of albumin. From this he concludes that *the source of life is fire,* and that the conditions for life were satisfied at the very time that the earth was incandescent: *Das Leben entstammt also dem Feuer* . . . . . As to the molecule of albumin, it was in reality formed only while the world was cooling, when the combinations of cyanogen and the hydrocarbons had contact with the oxygen of the *water.*

Even today the sun generates the constituents of albumin within plants. This excludes any idea of spontaneous generation. The living molecule of albumin is endowed with the faculty of growth; it is always in the process of formation and has no fixed characteristic composition or chemical equivalence. Under the influence of the sun, direct or not, it grows, and every living being is a simple molecule of albumin derived from the one and original albumin molecule, developed at the beginning of the terrestrial world.

From another standpoint, considering albumin as the basis of protoplasm, Pflüger examines, so to say, its chemical development in the two conditions of organization and disorganization. In protoplasm that is forming there would be a living albumin in which nitrogen is fixed in the form of *cyanogen;* in protoplasm that is being destroyed, a dead albumin in which the nitrogen is fixed in the form of *ammonia.* The transition from life to death, that is to say, its incorporation into protoplasm and its separation therefrom is thus characterized, for albumin, by the displacement of the molecule of nitrogen, which moves from the carbon to the hydrogen, and the entrance of albumin into vital activity is characterized by a reverse move.

This is just about the state of our knowledge on the question of organic creation or synthesis. We see that it is still, as at the time of Lavoisier, a profound mystery. Nevertheless research and hypotheses accumulate and a day will come when enlightenment will emerge from this long and painful work.

We ought in conclusion to return to a question that we have already touched upon, and ask ourselves if the chemistry of the laboratory, that is ordinarily involved in these circumstances, is indeed comparable to the chemistry of living beings. Lavoisier and many others of his successors seemed to believe it, but we have often shown that this direct application of the chemistry of the laboratory to the phenomena of life is not legitimate. We have insisted many times on this idea that the laws of general chemistry cannot be violated in living beings, but that they have nevertheless special agents and apparatus within them* which it is necessary for the physiologist to recognize. Is it necessary to go further, and say that really there are special chemical forces in living beings, and return thereby with Bichat to distinguish between vital properties and chemical properties? The words of certain chemists, who could be called vitalists, would seem to have this implication, which is why I think it useful to explain myself on this subject.

The *Traité de Chimie Organique* of Liebig begins with this phrase: *Organic chemistry deals with the materials that are produced in the organs under the influence of the vital force, and the decompositions that they undergo under the influence of other substances.* What is the meaning of this vital force which fabricates particular chemical products? One is led to believe that in the mind of the author it is indeed a matter of a vital force capable of executing what chemical forces cannot do; Liebig, in a word, expresses himself like a vitalist, and in another passage of his *Lettres sur la Chimie,* speaking of intoxication, he says, *Thus the vital force is overcome by the chemical forces.* We do not admit of any vital executive force; we have explained ourselves at length on this subject. Nevertheless we admit that vital phenomena exist in living beings, along with chemical compounds that are peculiar to them. How then is their production to be understood?

The chemistry of the laboratory and the chemistry of the living body are subject to the same laws; there are not two chemistries; this Lavoisier said. Only the chemistry of the laboratory is carried out by means of agents and apparatus that the chemist has created while the chemistry of the living being is carried out by agents and apparatus that the organism has created. We have more than amply demonstrated the truth of this proposition regarding the agents of analysis or organic destruction. The chemist, for example, converts starch into sugar by means of an acid that he has made; he saponifies fatty bodies with caustic potash, with concentrated sulphuric acid, with superheated steam, all

*See my *Rapport sur la Physiologie Générale.* 1867, p. 222.

agents that he has himself created. The animal, as well as the germinating seed, convert starch into sugar without acid, with the help of a ferment (diastase) which is a product of the organism. Fat is saponified in the animal, within the intestine, without caustic potash, without superheated steam, but by means of pancreatic juice, which is a product of secretion, given off by a gland. Each laboratory thus has its own special agents, but the chemical phenomena are basically the same; the conversion of starch into sugar and the splitting of fat into fatty acid and glycerine take place in both cases by an identical chemical mechanism.

It must be the same for the phenomena of organic creation. The chemistry of the laboratory can carry out syntheses like living bodies, and already it has produced a great number. Chemists have made essences, oils, fats, and acids that living organisms make for themselves. But here too it can be affirmed that the agents of synthesis differ. Even though the agents for synthesis in living bodies are not yet known, they certainly exist. We have mentioned the various hypotheses that have been proposed on this subject; for our part we have been led, by facts that we shall reveal later on, to attribute a certain role, not only to the protoplasm, but also to the nucleus of the cells.

In a word, the chemist in his laboratory and the living organism in its mechanisms work in the same way, but each with its own tools. The chemist can make the products of the living being, but it will never make its tools because they are the very result of the organic morphology which as we shall soon see is out of the realm of chemistry properly speaking; and in this connection, it is no more possible for the chemist to make the simplest ferment than to make the entire living being.

In résumé, we see how obscure all these questions of syntheses and of vital creations will remain, despite all the efforts devoted to their study.

For ourselves, we do not think that anyone will ever succeed in the solution of these complex problems by attempting to apprehend them at their very origin. We believe on the contrary that it is by following the facts observable closest to us that we can climb step by step and succeed in reaching determinism over these fundamental phenomena.

Today it can be said that the synthesis of complex materials, of albuminoid substances, of fatty bodies, is completely unknown to us. The only one about which we have some definite ideas is the synthesis of starch or glycogen in animals.

It is on this example that we must base our ideas of vital chemistry, since, in addition, it is at present the best known; it might be said that it is only one that is localized.

III. GLYCOGENIC SYNTHESIS

The most general result of the studies that we have carried out on this subject is to have proven that animals and plants have, the one and the other, the faculty of creating amylaceous and saccharine immediate principles. We are thus no longer of the belief that the animal is absolutely subordinated to the plant. The animal and the plant form the immediate principles that are necessary for their respective nutritions.

This fact is in accord with the general principle we put forward at the beginning of our studies, namely that life is not opposed, but alike in the two kingdoms, that it necessarily includes two orders of phenomena, organic creation and organic destruction, and every being endowed with life, whether animal or plant, whether simple protoplasmic or complete, must necessarily possess them.

It was about thirty years ago that I was led to discover the glycogenic function in animals. I was not led to it by preconceived ideas, but on the contrary by the pure and simple observation of the facts. At that time, everyone believed in the exclusive formation of sugar in plants. I was beginning a scientific career, and naturally had the opinions of my times. Thus I did not want to destroy the theory of exclusive glycogenesis; I tried first to support it and extend it. I asked myself how this sugar in the food that plants provided for animals was burned and destroyed in their organism. Not contenting myself with the hypotheses that had been proposed on this subject founded on the balance of foodstuffs on entry and exit from the animal organism, I undertook a series of experiments in which I proposed to follow within the blood, until its disappearance, the sugar ingested via the digestive tract in animals.

From my first trials I was very surprised to find that the blood of the dog always contains sugar, whatever its diet, and as well also when they are fasting. The fact is so easy to determine that it is astonishing that it had not been seen sooner; this is solely because everyone was dominated by preconceived ideas from which it was necessary to disengage, and because on the other hand the investigators who preceded me had neglected to follow strictly the rules of the experimental method.

As early as 1832 Tiedemann had found that the starch in the food could be converted into sugar and enter the blood; he found glucose in the intestine and then in the blood of a dog that had ingested starchy materials. From this Tiedemann drew the conclusion, new at the time, that sugar is formed normally in the intestine by means of the digestion of the starches and can pass from there into the chyle and into the blood. But if this investigator discovered nothing more, it is because in his experiments he neglected one of the most important precepts of the experimental method; he omitted controls. He contented himself, in fact, with saying that the sugar of the blood came from the ingested starch, but failed completely to investigate, in corroboration of his observation, whether the blood of animals that had not been fed starch were devoid of sugar.

It is this control that I carried out, and it is this that taught me that the blood of animals normally contains sugar, independent of the nature of the diet.

I went further and showed that in adult mammals it is in the liver that the formation of sugar takes place. The blood leaving the liver is always more abundantly supplied with sugar than that of any other part of the body.

After this discovery, an explanation was sought for the origin of the sugar in the tissue of the liver. At first cleavage and decomposition were thought of. Schmidt believed in a cleavage of fatty material giving rise to sugar in the blood. Lehmann believes that the fibrin of the blood, traversing the liver, splits into glucose on the one hand and bile acids on the other; Frerichs gives a similar explanation. Berthelot was tempted to believe in a cleavage within the liver of a material analogous to an amide; and I myself pursued experiments with this view in mind for some time.

I finally found that the material that is the generator of sugar within the liver is a true animal starch, *glycogen,* and in this way I was able to establish that the mode of formation of sugar is identical in the two kingdoms.*

Thus, in animals as well as in plants, sugar is formed at the expense of the starch. The formation of this starch in these two kingdoms is considered as an act of organic creation, a synthesis. The formation of sugar is on the contrary an organic destruction, a hydration of the starch which brings about its conversion into dextrine and into glucose; then this substance itself gives rise to lactic acid and to carbonic acid by a series of operations which have as a result the destruction of the sugar by procedures equivalent to the phenomena of oxidation.

Thus we find in glycogenesis in the animal, as in the glycogenesis in plants, the two characteristic phases of the great phenomena of life:

1. *Organic creation:* synthesis of starch, synthesis of glycogen
2. *Organic destruction:* conversion of the starch or the glycogen into dextrin and sugar, then destruction of the sugar by processes analogous to combustion.

Unfortunately we are not very well acquainted up to now with anything but the phenomena of destruction of the starchy principles; we know that in animals as in plants they take place under the influence of *ferments;* diastase, the lactic ferment, chemical agents peculiar to the organism. We know in addition that in the two kingdoms these phenomena generate heat when they take place.

As to the creation or the synthesis of starch or glycogen, it is for us surrounded by the greatest obscurity, in plants as well as in animals. At any rate we are making progress in the right direction, and it is probably in the animal that this formative mechanism will first be uncovered. I have carried out a great many experiments on this subject in mammals; their complexity makes them all difficult. By using fly larvae (maggots) I hope to be in a better position to discover the mechanism that gives rise to the very abundant glycogen in these larvae.

---

*See the summary of my *Recherches sur les glycogènes (Annales de Chimie et de Physique.* 1876).

To give an understanding of the difficulties of such studies on animals I would recall here the important fact that vivisection quickly disturbs and stops the phenomena of glycogenic synthesis, while it does not stop, and even in some cases accelerates the phenomena of destruction or transformation. This is why we have not yet been able to study, *post mortem,* by the processes of artificial analysis, anything but the phenomenon of glycogenic destruction while the corresponding phenomena of synthesis, like all phenomena of organic creation, for that matter, seem to require the integrity of the entire organism for their accomplishment.

Nevertheless the glycogenic material in animals as well as in plants is not destined only to be converted into sugar; it seems also to be made to enter directly into the constitution of the tissues during embryonic development.*

Whatever the role it has to play in the organism, the glycogenic material manifests itself to us in developing parts as the result of a true synthesis. The agent of this synthesis is the protoplasm of a cell. This cell, capable of producing glycogen, is found in the liver in the adult; it is variously located in the embryo, in the blastoderm, in the umbilical vesicle in the chick, and in the amnion in the ruminant, but it is likely that everywhere it produces the amylaceous material by the same process.

The glycogenic substance takes the form of granules or droplets, enclosed in the interior of the hepatic cells in the liver, in the cells of the blastoderm in the hen's egg, in muscle fibers in the fetus, and in epithelial tissue; it exists diffusely in a large number of embryonic tissues. During fetal life glycogenic cells are encountered in the placenta and on the vessels of the allantois (see Fig. 9*) .

The most interesting case is provided by the ruminants I have shown that in these animals the complete development of the glycogenic material can in fact be followed in its two phases of *formative synthesis* and organic *destruction.*

The glycogenic cells in the form of plaques (Figs. 10 and 11) accompany the allantoic vessels that here are reflected at random

*See *Compt. rend. de l'Académie des Sciences.* vol. XLVIII, 1859.

*See my memoir: *Sur une nouvelle fonction du placenta* (*Compt. rendus de l'Académie des Sciences.* vol. XLVIII, session of 10 January 1859.)

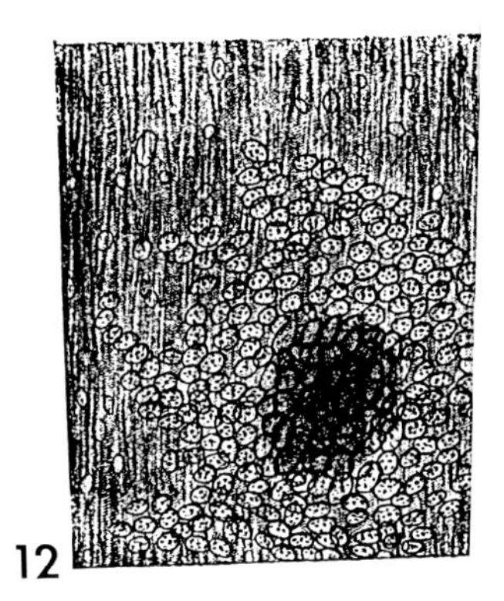

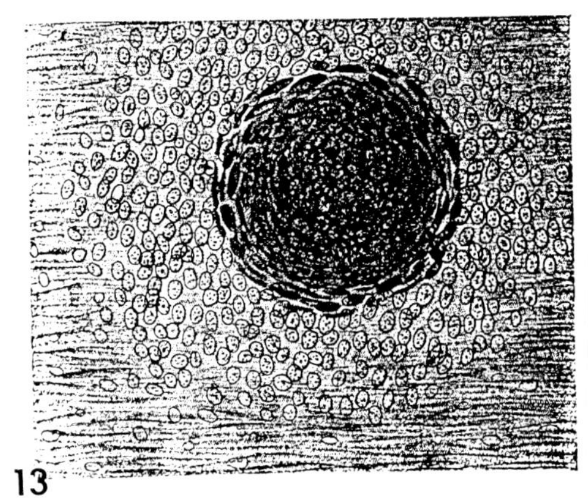

Figure 12, 13, 14—Beginning of the formatio
the amnion in a calf fetus.
Figure 12.—First stage: the small central mas
colored violet red by acidulated iodine wa
membrane are colored yellow by the iodine.
Figure 13.—More advanced stage: the mass o
a red color is more considerable.
Figure 14.—Glycogenic cells dissociated and

that produces heat and so contributes
the organism.

This example shows us vividly the
iate principle; its synthetic formation
cellular agent and then its destruction

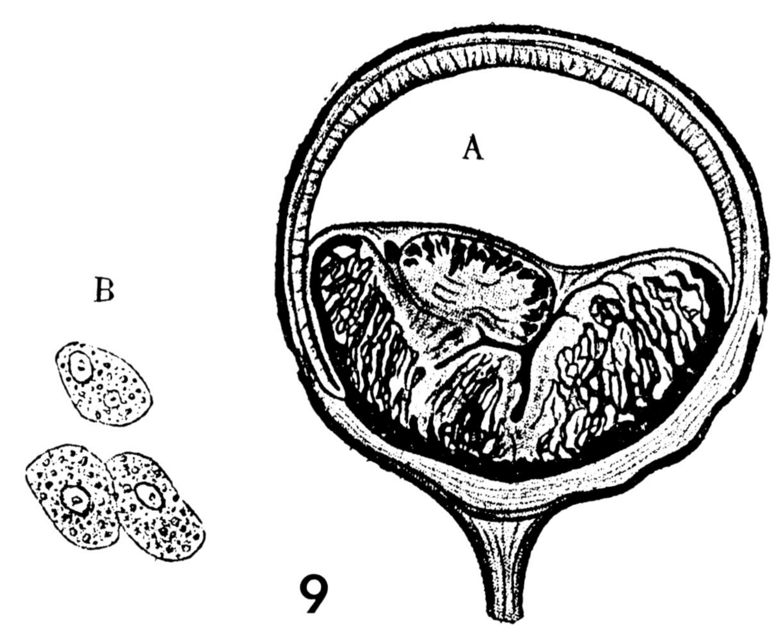

Figure 9.—Disposition of glycogenic cells in the placenta of the rabbit. A, Section of uterine horn and placenta *in situ*. The glycogenic cells are situated between the fetal placenta and the maternal placenta on the villosities of the allantoic vessels. B, Glycogenic cells of the placenta isolated and colored wine red by iodine.

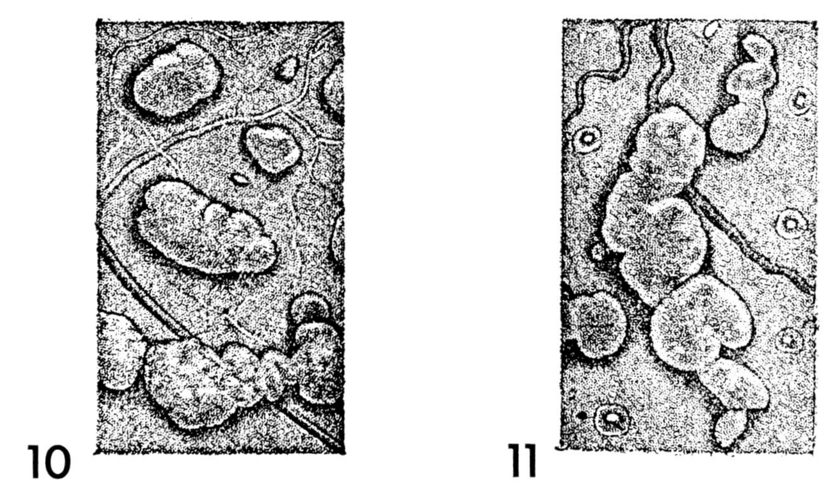

Figure 10 and 11.—Glycogenic plaques from the amnion of the calf fetus at the height of their development.

over the amnion. The glycoger
nants appear in the form of c
very beginning of embryonic l
gestation, then begin to be d
end of intrauterine life. The
thus measured by a span of ti
The plaques developed on th
whose transparency they beclo
as they grow; they cluster at co
(see Fig. 10). At the height of
present a thickness of several
culminating point separating
iod of destruction.

We have illustrated the va
ment in the amniotic plaques
14). The preparations (Figs.
phase of glycogenic developme
trates the culminating point
tions (Figs. 16, 17 and 18)
development.

The formation of glycoge
histologically as closely as it
tends to indicate that it takes
that of the production of epitl

The destruction of the pla
by resorption *in situ* or by re
which they fall. The plaque
pearance, and floats in the an
generation becomes more ma
the granulations disappear, a
the glycogenic material; oily
sometimes large crystals, and
fat, which is found at the bi
struction by oxidation takes
cium oxalate (Fig. 18) accur
the combustion that has take
was built up only to be torn

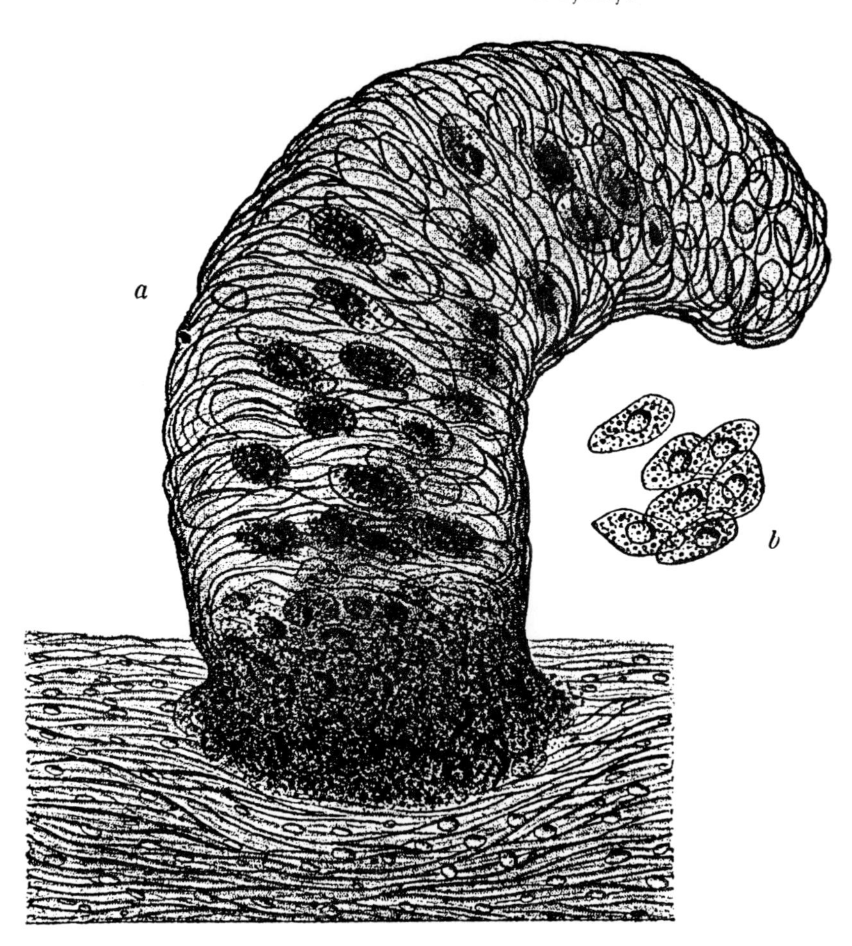

Figure 15. *a,* an isolated villosity from the glycogenic plaques. One sees certain cells which have been colored by the iodine, better outlined. *b,* cells of the villosity isolated and colored wine red by the iodine.

If we pursue the formation of the glycogenic material into the organs of the fetus,* we see that the glycogenic cells form in all the epithelia at the surface of the skin in the corneous tissues, beak,

*See my memoir: *De la matière glycogène considérée comme condition de développement de certains tissus chez le foetus avant l'apparition de la fonction glycogénique du foie (Comptes rendus de l'Académie des sciences,* vol. XLVIII, session of 4 April 1859)

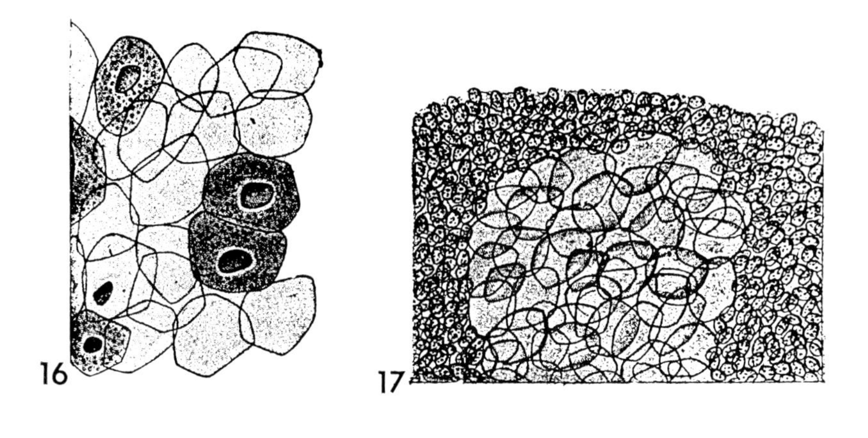

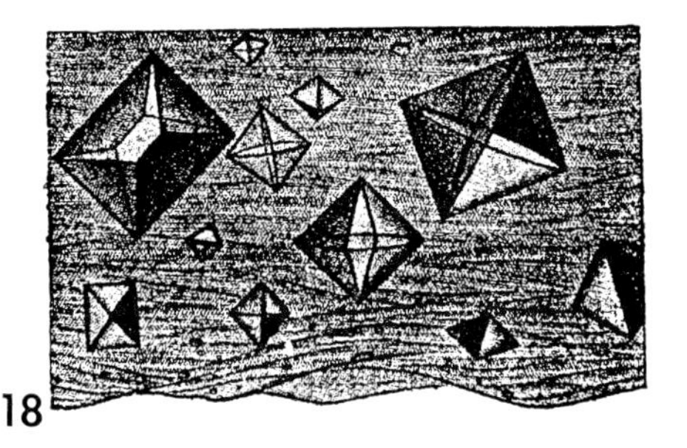

Figure 16, 17 and 18.—Degeneration of the plaques of the amnion of the calf fetus.

Figure 16.—Mixture of normal cells still retaining their nucleus and glycogen, and degenerated cells losing their nucleus, and no longer containing the glycogenic material, and proceeeding to fatty degeneration.

Figure 17.—Fatty degeneration of the glycogenic cells is complete.

Figure 18.—The glycogenic plaque has disappeared and in the place it occupied, one frequently finds various debris and crystals of calcium oxalate.

feathers, and toenails; in the epithelium of the intestine and the lungs, in the glandular ducts; but never in the glandular tissue itself, nor in the lymph nodes, nor in the *endothelia,* etc., etc.

What is curious is that the liver which in the adult will become the site of choice for the formation of glycogen, yet contains no trace of this substance. In the calf it is about at the middle of gestation that the liver acquires this property, and then the glycogenic material is seen to disappear from the epithelia, and the glycogenic function ceases to be diffuse, to be localized in the liver.

In the lower creatures which have no liver, the glycogenic function always remain diffuse, as in plants.

In certain animals, such as the crustacea, this function is intermittent and corresponds to period of moulting, as it corresponds to growth in plants, etc., etc.

The protoplasm of the cell is necessary only for the first phase, that is to say for the synthetic genesis of the immediate product, but the destructive combustion can take place without intervention by the protoplasm. Proofs of this abound. The glycogenic material is an example of this; nothing can substitute in its production for animal or plant protoplasm; on the contrary its destruction is a chemical phenomenon which does not necessarily require the intervention of the living cellular agent, and can continue after death or outside the body, A decisive experiment in this regard is that of the *perfused liver.* A current of fresh water is passed through a liver removed from the animal's body and in consequence removed from any vital influence; in this way all the sugarlike material it contained is removed. After the organ is set aside for a time a new quantity of sugar is again found. The test can be repeated a great number of times with the same success, until the stores of the glycogenic material are exhausted. Thus in this dead organ, isolated from all physiological or vital influences, the glycogenic material continues to be destroyed as in life, but it is not reconstituted.

How does the cellular protoplasm intervene to form the immediate principle? This is a question to be resolved. Perhaps it could be supposed that the glycogen appears, not by a true synthesis in the chemical sense of the word, but by a cleavage of the protoplasmic material. It belongs to the future, and probably to a near future, to resolve these problems that can only be pointed out today, but whose principal conditions we have already succeeded in analyzing.

SEVENTH LECTURE

Properties of Protoplasm in the Two Realms
Irritability, Sensibility

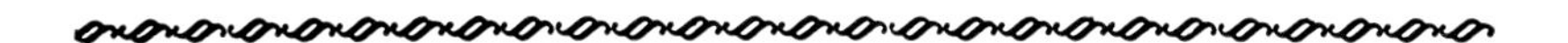

SUMMARY: Protoplasm possesses irritability and motility—These properties constitute the connection between the organism and the external world.

I. History of irritability—Glisson, Barthez, Bordeu, Haller, Broussais, Virchow—Irritability; autonomy of the tissues—Protoplasm is the seat of irritability.

II. Excitants and Anesthetics of Irritability—Normal conditions for protoplasmic irritability—Anesthesia* of protoplasmic properties of the activity of irritability or sensibility in animals and plants—Experiments—Anesthesia of the protoplasmic phenomena of germination, development, and fermentation in animals and plants—Anesthesia of the germination of seeds—Anesthesia of eggs—Anesthesia of organized ferments—The impossibility of anesthetizing soluble ferments—Anesthesia of the chlorophyllian function in plants—Anesthesia of the worms of rusty wheat.

III. Irritability and Sensibility—Conscious and unconscious sensibility—Different points of view of philosophers and physiologists on this subject—Identity of the anesthetic agents abolishing sensibility and irritability—We do not act on the properties or functions of nerves, but solely on the protoplasm.

*The word *anesthesia* designates here the action of the anesthetic agents, ether or chloroform, leading to the suppression of the ability of the elements and of the tissues to react under the influence of their ordinary excitants.

THE PROTOPLASM, THE AGENT of the phenomena of organic creation, does not possess within itself only the power of chemical synthesis that we have examined; to put this power into operation it must have the faculties of *irritability* and *motility*. It must in fact react and contract under the impetus of excitants from the outside, for it does not have within itself or by itself any initiative faculty.

The phenomena of life are not the spontaneous manifestation of an internal vital principle; they are on the contrary, as we have said, the result of a conflict between the living material and the external conditions. Life occurs from the constant reciprocal relations between these two factors, both in the manifestations of sensibility and movement, which are usually considered as belonging to the highest orders, and in those related to physicochemical phenomena.

This constant interaction between the organized substance and the ambient environment is thus a general characteristic of organic life as well as animal life. Nutrition, as well as sensibility and movement, express in more or less complicated forms this faculty of living matter to react to excitation from the external world. This faculty, the essential condition for all the phenomena of life, in the plant as well as the animal, exists in its simplest form in the protoplasm. This is *irritability*.

In a general way, *irritability is the property possessed by every anatomical element (that is to say, the protoplasm that enters into its composition) to be set into action and to react in a certain way under the influence of external excitants.*

Every manifestation of life requiring the concurrence of certain external conditions or excitants is by that very fact a manifestation of *irritability*. Sensibility, which at its highest level is a complex phenomenon, is basically, as we shall see, only a special modality of irritability, the only elementary vital phenomena whose existence is common to the two kingdoms.

We must first examine what is meant by this word *irritability*, and understand what ideas and what facts it denotes. It is necessary to know the historical antecedants of this fundamental question which for more than a century has given rise to continual confusion and has started controversies that are not yet at an end.

The problem of the sensibility of living beings, and, in a general way that of the *vital properties* of organized beings, will find their solution in the understanding and the exact evaluation of the doctrine of *irritability*.

I. HISTORY

It was Glisson (1634-1677), Professor at the University of Cambridge, who first introduced *irritability* into physiological explanations as a vital property which he attributed to all "animal fibers, muscular or otherwise," that is to say, nonspecifically to all organized matter; to him it was the cause of life.

From the moment this expression was employed, it has given rise to endless confusion; the three properties and the three terms, namely, *sensibility, irritability,* and *contractility,* have been distinguished and confused and again separated and again identified. From this have arisen misunderstandings that must be dissipated.

Barthez (1734), the creator of the doctrine of vitalism, distinguished *sensitive forces,* or sensibility with perception and sensibility without perception, and *motor forces,* of contraction, elongation, fixation and tonus, equivalents of present-day *contractility;* these two kinds of forces were moreover subordinate in the living being to the *vital force*.

It is said that Leibnitz accepted Glisson's doctrine of irritability; the perceptive entelechy that he considered to be the principle of action inseparable from living particles would be nothing but *irritability* under another name. The relations of Leibnitz with Campanella and Glisson would permit the supposition that this interpretation might have presented itself to the mind of the great philosopher.

Bordeu (1742) distinguished a single vital property, *general sensibility,* which included them all. This is the earliest source of the confusion that we have mentioned! Bordeu used this word in a new and unusual sense. He designated by it what in his time was called *irritation, excitation,* the *irritability* of Glisson, the *incitability* of *Brown,* that is to say the property of reacting under the influence of a stimulus, to which the English physician Brown (1735-1798) had attached so much importance.

The innovation of Bordeu is to have generalized *sensibility* to the point (as Cuvier reproached him) of giving this name to "all nervous interactions accompanied by movement, even when the animal had no perception of it."

In addition to this general sensibility, the basis of which is the same for all parts, Bordeu conceived also of an *intrinsic sensibility* for each of the parts: "Each gland, each nerve has its particular style. Each organized part of the living body has its way of existing, of acting, of feeling and moving, each has its style, its structure, its internal and external form, its weight, its very particular way of growing, of extending and retracting; each contributes in its own way and its own share to the ensemble of all the functions, and to the *general life;* finally, each has its own life and its own functions distinct from all the others." Bordeu went as far as to say that *"every organ* is an animal within the animal: *animal in animali,"* a doctrinal excess that aroused the criticism of Cuvier and more recently of Flourens.

This is the point of view of Bordeu relative to the vital properties or *special sensibilities.*

It was Haller, the celebrated physiologist of Lausanne, who had the honor of providing an experimental basis for the theory of the vital properties, and establishing it solidly. He distinguishes three properties:

1. *Contractility,* which is nothing more than the physical property that we call *elasticity* today.

2. *Irritability,* just as poorly named. It is the way muscle behaves. Hallerian irritability is the *contractility* of today. Muscles, said Haller, are *irritable,* now one says *contractile.*

3. *Sensibility.* This is the way nerves behave.

It can be seen by this that the distinction as established by Haller has a practical and experimental character. He does not concern himself with the essence of the properties he establishes. He sees nerves and muscles behaving differently, and he gives different names to these two kinds of activity: irritability and sensibility. The result of his experiments was thus to separate (which had not been done before him) nerve and muscle, from the standpoint of their mode of action, and to separate the one and the other from different tissues, tendons, epidermis, cartilage, which behave otherwise.

It is the principal merit of Haller to have shown that the nerve and the muscle have within themselves whatever is necessary for them to enter into action, and that they do not derive their activating principle from elsewhere. The doctrine prevailing since Galen and accepted by Descartes, the doctrine of animal spirits, taught that the organs received their activating principle from a central force transmitted and distributed by the nerves under the name of animal spirits, and led in fact to the supposition that the muscle derived the property of contracting from the nerve. Before refuting this accepted error experimentally, and to demonstrate the autonomy of the two tissues and their independence by direct proofs. Haller established ingeniously and *a priori* the lack of basis for the current doctrine. He observed that if the muscle were to derive its properties from the nerve, the number of nerves that activated a muscle ought to be proportional to its volume, a conclusion that is not in accord with the facts; the heart for example, which is the most active muscle in the body, is the one whose innervation is the least abundant and the most difficult to demonstrate.

The demonstration of the essential independence of nerve and muscle, attempted by Haller, was completed later by J. Müller, who proved that a nerve separated from the body is disabled before the muscle. The principles of the activity of the two tissues cannot be the same, since the one has disappeared while the other persists. As to the objections to which Müller's argument was liable, I refuted them later in my experiments on curare, which suppresses the action of the nerve completely while permitting the activity of the muscle to survive *in toto*. Here we ought to add a reflection; curare destroys a mechanism; its action does not bear upon the protoplasm, that is to say, on the physical basis of the life of the tissue. Curare destroys the *physical relationship* of nerve and muscle; an indispensable relationship for carrying out voluntary contraction and voluntary movement. It separates elements normally united, it destroys their harmony, while at the same time not destroying the elements themselves.

In résumé, all these investigations undertaken with irritability in mind have ended by proving the *autonomy* of the tissues; they have not clarified the question of irritability, which has remained where it was. The property of nerves called sensibility or motricity and the property of muscles called contractility are by no means general attributes of all living matter, but rather reactions, or special manifestations of a particular kind of living matter. These are special properties and not general properties of life. When the fundamentals of things are examined carefully it can be seen that these properties are only particular expressions of a more general property, *irritability*.

This is what Broussais thought.

Broussais accepted only one essential property of organized material, *irritability,* involving as a consequence, sensibility, contractility, and all the other secondary faculties. Virchow professed the same opinion; the vital phenomena have *irritability* as their intimate cause, a generic term that includes, according to him, *nutritive irritability, formative irritability,* and *functional irritability*.

By the word *irritability* Virchow referred to "the property of living bodies which makes them capable of passing into the active state under the influence of irritants, that is to say, external agents."

In other words, we would say, for ourselves, that "irritability is the property of the living element to act according to its nature under outside provocation." First of all each tissue reacts to the excitation of the external environment of water, air, heat, and food, by taking certain principles from it and giving off others to it, that is to say, by carrying out the exchanges that constitute nutrition. It is this that is called *nutritive irritability,* or the property of reacting to the stimulus of the foodstuffs in the ambient environment by taking nourishment from it. In addition, each element has the opportunity to manifest its particular properties and to act in a special and characteristic way: muscle fiber reacts by contracting; nerve fibers, by conducting the disturbances they receive; glandular cells, by elaborating and discharging special secretory products; vibratile cilia, by flexing and extending alternately; blood corpuscles, by attracting oxygen; and chlorophyll granules, by decomposing carbonic acid. These are all faculties that have been designated by the generic term *functional irritability*. But all these particular manifestations are dominated by a general condition; they are various kinds of a single faculty, simple *irritability*. It is not necessary, according to us, to distinguish between nutritive irritability and functional irritability, still less is it necessary to establish distinctions within each of these properties, and like Virchow dismember nutritive irritability into a *formative irritability* which would be the property of a tissue to maintain itself by the generation of cells or anatomic elements that would succeed each other, and into an *aggregational irritability,* a property of the elements of incorporating appropriate foodstuffs. It is basically the same essential property that characterizes the relations between organized and living substances, or *protoplasm,* on the one hand, and the external environment on the other; the simplest and the most general faculty of life, in animals as well as in plants is irritability.

The innumerable experimental studies that have been carried out on the properties of living tissues, and which we cannot retrace here, lead to these two conclusions:

1. There is in all living tissues a common faculty of reacting under the influence of external excitants; this is *irritability*. The tissue can be called living only on this condition.
2. There exists at the same time in all living tissues a particular and innate reaction, the *organic property,* which characterizes the tissue physiologically.

Now, in what constituent part of the tissue ought we to localize these two properties, the one which is common to all, and the other which is peculiar to each?

It is in the protoplasm alone that we find the explanation of all the properties of the tissue. In reality, the protoplasm possesses all the vital properties, in a more or less comingled state; it is the agent of all organic syntheses, and by this very fact of all the intimate phenomena of nutrition. Moreover, the protoplasm moves and contracts under the influence of excitants, and thus presides over animal life.

In the development of the organisms and the successive differentiation of their tissues each of these primary and interrelated properties of the protoplasm is itself differentiated to a relative degree that is greater in certain organic elements. Thus the autonomy of tissues is basically nothing more than a protoplasmic differentiation. At any rate, in each tissue, whatever speciality it assumes, the protoplasm never loses the faculty of feeling the excitants that must enter into contact or into conflict with it to produce the manifestation of one of its special properties. In certain cells the external irritation produces the synthesis of certain ternary and quaternary materials, in the form of solid or liquid secretions; at one moment it is the synthetic property of the protoplasm that has been set into action; at other times, the external irritation induces a multiplication of cells and set into activity the proliferating property of the protoplasm; at other times, finally, the external irritant will excite muscular contraction and manifest the motor or contractile property of the protoplasm.

Such is the concept we ought to form of protoplasm; it is the origin of everything, it is the sole living material of the body that animates all the others. It is a part of the ancestral protoplasm that develops the new being, it is by the incessant reproduction of the protoplasm that life is perpetuated.

We shall not give an account here of all the properties of protoplasm; this would embrace the whole of physiology. In what follows we shall concern ourselves only with its dominant property, sensibility or irritability, without which the rest are as nothing, and remain incapable of manifestation. We shall direct our study most particularly to the action of excitants and anesthetics on protoplasmic irritability.

II. EXCITANTS AND ANESTHETICS OF IRRITABILITY

The conditions for the release of irritability are known; we have examined them in the study of latent life, for, it must be kept in mind, latent life cannot be terminated unless the protoplasm in some way reawakens, that is to say, resumes its property of irritability. The excitants of irritability are thus those of life itself; they are water, heat, oxygen, and certain substances dissolved in the ambient medium.

Without doubt the extrinsic conditions that must be provided to permit the protoplasm of each cell to live and function according to its own nature are very numerous, very variable, and very delicate. If they were to be specified in all their details, since the nature of the excitants, their dosage, and their varieties are infinite, to know them it would be necessary to give an account of each particular cellular element.

But to restrict ourselves to general and essential conditions, we shall say that they are the same for every kind of protoplasm, animal or plant; they are the four conditions that we have indicated above.

By a singular coincidence, it can be said that these four indispensable conditions for the exercise of irritability and of life are precisely the four elements that the ancients considered to make up the world; water, air, fire (heat), and earth (chemical substances, nutritive or saline), which the living being encounters in the ambient medium.

Relative to the physicochemical conditions for life, we have nothing essential to add to what we have already said in a general way about the conditions for latent life, oscillating life, and manifest life.

We shall on the contrary spend some time on the action of anesthetics on irritability, on which we have carried out special studies on animals and on plants.

The anesthetics, ether and chloroform, afford us the means of acting upon irritability, the vital faculty par excellence, and to suspend it or suppress it, so that these substances can be thought of as *natural reagents acting on all living substances, and in consequence on protoplasm.*

These substances have the faculty of suspending the activity of protoplasm, of whatever kind it might be, and in whatever way it is manifested. All the phenomena which are truly dependent upon *vital irritability* are suspended or suppressed permanently; the other phenomena, of a purely chemical nature, which take place in the living organism without the aid of irritability, are on the contrary preserved. From this derives an extremely valuable method of distinguishing in the manifestations of the living being that which is *vital* from that which is not.

These views are not purely theoretical; they are on the contrary suggested and demonstrated by experiments that we have recently begun, and which we shall permit you to witness in their turn.

Everyone knows that the anesthetics, ether and chloroform, have the property of temporarily extinguishing sensibility, and in consequence keep the patient operated on from being conscious of or remembering pain, which is the same as its abolition. But we have found that this action of the anesthetics is general, that it does not apply only to the phenomenon of consciousness, that is called pain or sensibility, but that it acts upon the *irritability of the protoplasm,* and extends to all vital manifestations, of whatever kind they might be. This ought to be so, since we refer all vital activities to the protoplasm.

The action of anesthetics is expressed by a more or less rapid effect on various organisms and on their various tissues. The first point on which it is necessary to insist is that the etherizing action extends successively to all the tissues in the same being. When a man is anesthetized, for example, by means of chloroform or ether, the anesthetic substance is inhaled, absorbed by the lungs, and circulated to the tissues in the blood. It is on the more delicate protoplasm of the nervous centers that the anesthesia first exerts its action and it is in fact the phenomena of consciousness and sensory perception that are the first to disappear; while the protoplasm of the nerves, the muscles, the glands, and other anatomical elements is not yet affected. This explains why the vital functions can continue to be exercised and why anesthesia is therefore without danger to life, for if the protoplasms of all the anatomical elements in all the tissues were affected at the same moment by the anesthesia, all functions would stop simultaneously, and death would be instantaneous. Surgical anesthesia is thus an essentially incomplete anesthesia; it affects only the most delicate nerve elements, which are the seat of the phenomena of conscious sensibility, and this suffices for the purpose in mind. But here we wish to demonstrate that anesthesia is a general phenomenon in all tissues, and we ought to demonstrate it both in animals and in plants.

Phenomena of Anesthesia of Movement and Sensibility in Animals and in Plants

The influence of anesthetics can be studied in animals and in plants as well. Many plants in fact exhibit phenomena of motor reactions in close relation to external stimulation, like the manifestations of sensibility in animals. Examples of movement appropriate to some purpose abound in the cryptogametes.

It is known that at the frontier between the two kingdoms there is a whole group of disputed beings that cannot be annexed to either of the two. The plant amoebas the *Plasmodia* studied by de Bary exhibit a mixture of the traits of animals and plants. These are protoplasmic masses that do not form cells or tissues during their whole period of development, they move by creeping over the debris of decomposing plants, on bark, on tan-bark. They send out prolongations, sorts of arms, in which the granular protoplasmic material accumulates. Their structural appearance, organization, and mode of reptation establish the closest analogies between these plant myxomycetes and the protistan animals of Haeckel.

The faculty of movement is encountered most clearly and evidently in the reproductive apparatus of algae, the zoospores. These are little egg-shaped masses ending in a breech or rostrum furnished with two to four cilia. These corpuscles move, change their location, and direct their course while swimming; they seem often to avoid obstacles, to make several trials to get around them, and get to a determined location. Here is to be found not only simple movement, but movement appropriate to a determined goal, the semblance, in a word, of voluntary motion.

The characters of voluntary movement are encountered even more clearly in the antherozoids of certain algae, the Oedogonia, for example. Pringsheim saw these antherozoids in 1854, as male reproductive bodies, in the form of a wedge, with a rostrum armed with cilia. Once it leaves the cell that contained it, the antherozoid swims in the surrounding water and directs itself toward the female cell; it comes in contact with the wall of this cell and seeks the orifice that this latter presents. After several fruitless attempts, it appears that a better directed effort permits it to gain the narrow canal and to embed itself in the green matter of the oosphere, the cell where fertilization takes place.

These examples of movements are not rare among the phanerogamic plants. The number of plants whose leaf organs are capable of movement is very considerable. Some of these movements are provoked by touching and shaking; others by the action of light and heat; still others seem to take place spontaneously through the action of internal causes.

We shall cite in particular the movements of the stamen of the spiny barbery *(Berberis)*, of the sundew or drosera, the fly-trap *(Dionaea muscipula)*, and the gyrating sainfoin *(Hedysarum gyrans)*.

The primary condition for these manifestations of movement is the faculty of reacting to the external excitants that provoke them; this faculty is not the exclusive property of animals. Many plants are endowed with it to a more or less pronounced degree.

The legumes belonging to the genera *Smithia, Aeschynomena, Desmanthus, Robinia,* our false acacia; the *Oxalis sensitiva* of India, exhibit the remarkable faculty of reacting to excitations applied to them. But the most celebrated species in this respect, and the most studied, is the Sensitive plant, *Mimosa pudica.*

The leaves of the Sensitive plant are arranged like compound pennate leaves, on four secondary petioles, themselves carried by a common stem (see Figs. 19, 20). When the plant is subjected to an excitant of any kind, the common stem lowers, the secondary petioles come together, and the leaves fold against each other on the upper surface. The irritation extends more or less according to its strength. It can be produced by most agents that are recognized as excitants of sensibility in animals, such as blows, shocks, burns, the action of caustic substances, and electrical discharge. It seems that some of these excitants become weaker with use or from fatigue. There is a sort of habituation which causes the plant to respond to stimulation with less and less strength, the more they are repeated. The naturalist Desfontaines observed this fact while transporting a Sensitive plant. The first jolts of the carriage caused the leaves to fold up and the stems to lower. But soon the leaves recovered and expanded again. Repeated starts and stops produced a repetition of the same phenomena at a constantly diminishing intensity.

Figure 19. Sensitive plant *(Mimosa pudica)* placed in an atmosphere of ether—*e,* sponge soaked in ether; The leaves of the plant are spread out, have become insensitive, and no longer close when they are touched.

We have spoken above of the practice that is well known in surgery today under the name of *anesthesia.* The agents that can be used to make man and animals insensible are ether and chloroform. Very well! It is a most unusual fact that plants can be anesthetized like animals, and all the phenomena can be observed in absolutely the same manner. Here there have been placed, each under a separate bell-jar, a bird, a mouse a frog and a sensitive plant. Inside each jar a sponge soaked in ether is introduced. The anesthetizing influence does not delay in making itself felt; it follows the level of the organisms. It is the bird, which has the highest organization, that is the first to be affected; it staggers and falls insensible after four to five minutes. It is then the turn of the mouse; after ten minutes it can be excited and its paw or tail pinched without movement. It is completely insensible and does not react. The frog is paralyzed later, and you see it after it was removed from the bell-jar, limp and indifferent to external excitations. Finally, the sensitive plant remains. It is only after twenty-five minutes that insensibility begins to be manifest.

We have placed under the bell-jar *C* (Fig. 19) a very active sensitive plant. At the side of the pot a wet sponge *e*, impregnated with ether is placed. Soon the ether vapor fills the bell-jar and acts on the plant. The anesthetizing action is more rapid in warm weather than in cold weather, and follows the various circumstances that increase or decrease the irritability of the sensitive plant. It is therefore necessary to vary the quantity of anesthetic according to these diverse circumstances. Here we are working in the shade, with diffuse light; if we were to work in the sun the effect would be much more prompt, but also more dangerous; often in this case the plant is killed and does not again recover its sensibility. This singular and special influence of the light from the sun that we confirm here regarding the action of ether or chloroform on the sensitive plant, we will encounter again later in many other phenomena of plant life.

Now after about a half an hour the sensitive plant is anesthetized, and we see that touching its leaflets no longer causes their lowering, while the same excitation produces an immediate contraction of the folioles *f* in a normal sensitive plant *(see* Fig. 20). Again we observe the fact that anesthesia first affects the pads of the folioles and then the pads placed at the base of the common stem of the compound leaf.

Some time has now passed, and you see that the anesthetized sparrow, white rat, and frog have now regained their sensibility and their movement; soon it will be the same for our sensitive plant; it will cease to be under the influence of the ether and regain its sensibility as before.

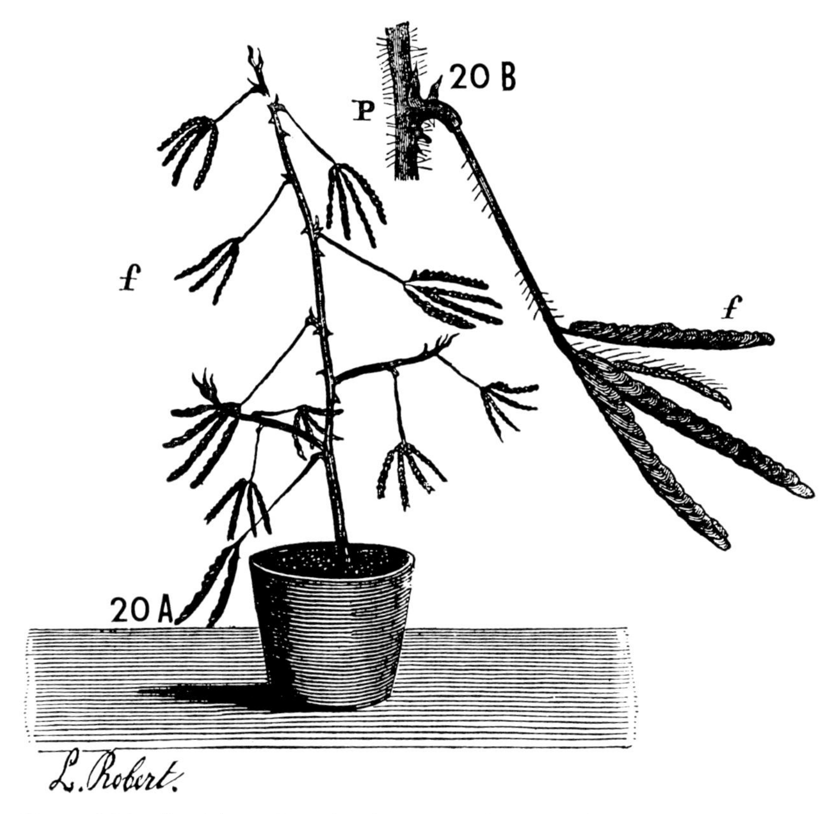

Figure 20A.—Sensitive plant in the state of contraction. Its leaves are retracted and lowered under the influence of a mechanical excitation applied to the plant.
Figure 20B.—Isolated leaf of the sensitive plant, to show the swelling at the base of the petiole and where the site of the contractile plant tissue is located.

The result of anesthesia is thus the same in animals and in plants. What we see here for the sensitive plant is true in fact for all the other movements that we have noticed in plants; movements of the spines in the spiny barberry, etc. It remains to be determined whether the mechanism by which this phenomenon is produced is identical. This is a most important question to resolve. If the analogy of the effects holds as far as the mode of action, it can be understood what an intimate relationship would thus be manifested between the animal organization and the plant organization.

First let us recall how ether or chloroform act upon the animal.

In anesthesia in man and animals, as it is ordinarily practiced, the anesthetic agent comes in contact with the lungs or skin along with the respired air; it is absorbed, enters the blood and perfuses all the organs, all the tissues and the anatomic elements. The action of the anesthetic substance is ordinarily explained by saying that of all the anatomic elements with which it is placed in contact, only one, peculiar to the animal, is affected; that is, the sensitive element, the cerebral element of the central nervous system. From this it results that sensibility is destroyed in the seat of perception, and in consequence pain abolished.

If this interpretation were true, the experiments that we have just carried out before you would remain incomprehensible to you, and no analogy could be established between the animal and the plant. For in the plant no nervous system is to be found, no central nervous center, no brain. It is indeed true that some authors, Dutrochet himself, have thought that they found in the sensibility of plants the proof that they had some organ analogous to the nerves, and there are even some (Leclerc of Tours) who have pushed the idea of systems, and of improbability, so far as to propose, in the sensitive plant, the existence of a nervous apparatus, of a cerebrum and cerebellum.

Some authors, the distinguished botanists Unger and Sachs of Würtzburg, consider the movements in question to be the result of the loss of equilibrium between two opposite forces, namely, the endosmotic attraction of the contents of the cells for external moisture, and the elasticity of the cellular membranes. But whatever the intimate mechanism of these phenomena might be, we cannot attribute their suppression to anything but the disappearance of the irritability of the contractile cells of the plant.

In fact, the anesthetic agent does not act exclusively on the nervous system, in reality it exerts its action on all the tissues of the animal; it affects every element, in its own time, according to its susceptibility. Just as it strikes the bird most rapidly and more slowly, the mouse, the frog, and the plant, according thus to the status of the creature, so in an animal organism it follows so to say the scale of the tissues. The effect shows up in the other systems after it has already been manifest in the nervous system, the most delicate of all. It is this that explains why the anesthetic influence on this element is the first in point of time.

Thus all tissues respond in the same way to the action of the anesthetic agent, in all there is the same essential property whose action is suspended; this property is the *irritability of the protoplasm.*

In résumé, the anesthetic agent affects the activity common to all the elements; it affects, suspends, or destroys the general irritability of their protoplasm. It causes the irritability to disappear temporarily if the contact lasts for a short time, permanently if it is prolonged. And this, we have seen, takes place everywhere irritability exists, in plants as well as in animals.

We have said that in our experiments the anesthetic agent did not act on sensibility as a function, but on the irritability of the protoplasm, as a property of the sensory nerve fiber or cell; thus the manifestation of sensibility and the expression of pain would find themselves suppressed, as well as the functional consequences which would result therefrom. What we say here is true not only for the irritability of the sensory nerve element, but for the irritability of the motor element and of all the living elements of the body.

The experimental proof is easy to provide.

Let us take for example the muscular tissue of the heart. Here is the heart of a frog, removed from the body, which continues to beat by the very virtue of its irritability, which persists: we place it in an etherized atmosphere. Soon its beats stop, to be resumed anew when we suspend the influence of the ether.

Let us take still another tissue, the ciliary epithelium, which moves incessantly by virtue of its irritability. The ciliary epithelium is easily observed in the frog's esophagus, where it forms the internal coat. The cilia that surmount the epithelial cells are animated by a constant motion which persists for a long time after the irritability of the other animal tissues has already been extinguished completely. By spreading out the membrane of the esophagus of the frog on a cork plate, as you see here, and placing on it some small grains of animal charcoal, one can see them carried along by the action of the cilia from the mouth to the stomach. The movement can be followed with the naked eye, and they can all be seen to go against the force of gravity. This action of the vibratile cilia of the esophageal membrane is powerful enough to carry quite heavy bodies, such as particles of lead, etc. The vibratile movements are well known and have been studied thoroughly.

They can be amplified by means of a very simple apparatus which makes them appreciable at a distance. You see such an apparatus. A glass slip rests on the membrane and is displaced, pulling a very long and very light lever made from a straw and capable of rotating at one end. The displacement of the lever thus makes us aware of the movements of the vibratile cilia.

What we want to demonstrate here is that the vapor of ether or chloroform stops the movement and lets the cilia fall into quiescence; then it can be noted that the transport of the particles on the surface of the esophageal membrane stops, to resume when the etherization has dissipated.

How is the irritability of the tissues or of the elements of the tissues affected by the ether? Evidently as the consequence of some chemical or molecular change that the anesthetic poison produces in the very substance of the element. From experiments that I have carried out at other times I think that this modification consists in a sort of coagulation. Ether coagulates the protoplasm of the nerve elements; it coagulates the contents of the muscle fiber, and produces a muscular rigidity similar to *rigor mortis*. In the physiological state the tissues and the elements of the tissues cannot manifest their activity except under conditions of humidity and semifluidity peculiar to their matter. Thus during life the muscular substance is semifluid; if this physical state ceases to exist, and if a coagulation takes place, function is suspended, as for example when water is frozen; its mechanical properties are lost until its fluid state returns. Finally we can add that these modifications in the physicochemical state of organized matter, although transitory, end by causing the death of the organism when they are repeated a certain number of times, because then, doubtless, the element does not have the time to recover sufficiently in the rest periods.

Direct experiment shows us this coagulation of the muscular element produced by the action of ether.* If a muscle is placed in ether vapor or if lightly etherized water is injected into the muscular tissue, after a certain time a definitive rigidity of the muscle is produced; the contents of the fiber are coagulated. But prior to this extreme state, there comes a time when the muscle has lost its irritability; it is anesthetized. If then the muscle fiber is observed under the microscope, it is seen that the contents are no longer transparent, that it is opaque and in a state of semicoagulation. These phenomena are seen very well in the frog by injecting the etherized water into the thickness of the gastrocnemius muscle; in this way we obtain a local anesthesia, a suspension of the irritability of the muscle, which no longer contracts. Letting the animal rest, we see the muscle return little by little to its normal state; the coagulation of its contents, and the rigidity disappear from the anatomic elements as they are perfused and irrigated constantly by the flow of blood.

It can be supposed that something similar takes place in the nerve.

Experiment makes it certain that ether and chloroform act indeed as natural reagents on all living substance, their action reveals in sensibility a common property of all living beings, animal or plants, simple or complex. In consequence, far from sensibility and motility being, as Linneus proposed, a character distinguishing between the two kingdoms, anesthetics on the contrary establish their relationship and their unity on a solid physiological basis, as their structural analogy has already established their vital unity in the anatomical sphere.

But it is not only upon the irritability of the protoplasm of the organic elements, sensory and motor, that anesthetic agents exert their action; they also affect the protoplasm of the organic elements which act in chemical synthesis, in the phenomena of

---

*Cl. Bernard, *Leçons sur les Anesthésiques et sur L'asphyxie*. Paris, 1875, p. 154.

germination and of fermentation, in a word, in the phenomena of nutrition.

Phenomena of Anesthesia of Protoplasm in Phenomena of Germination, of Development, of Nutrition, and of Fermentation in Animals and Plants

Anesthesia of Germination

Some years ago we established that ether or chloroform suspend the germination of seeds.

Here, as it might be said, germinative irritability is affected.

This is how we arrange the experiments: we take the seeds of *garden cress,* which germinate very quickly, and we place them in the necessary and sufficient conditions for their germination, air, humidity, and appropriate warmth, but at the same time in an anesthetizing atmosphere. We do everything in the same way with similar seeds placed in identical circumstances, minus the presence of the anesthetic agent.

In a preliminary experimental protocol (see Fig. 21) we caused, in comparison, a current of ordinary air and a current of air containing anesthetic vapors to pass over moist sponges and in two test tubes, containing seeds of garden cress on their surfaces.

A pump *P* attached to a faucet *R,* connected to the test tubes by rubber tubing *bb′* is designed to aspirate the test tubes and to make air pass through them. But in one case the test tube aspirates the outside air directly through tube *a′* connected at its lower end, while in the other case the air entering by the tube *a* must first pass through an initial test tube *t,* at the bottom of which there is a layer of ether *S.* Thus the air is loaded with this ether vapor which saturates the atmosphere inside the test tube, and is carried into the test tube by the rubber tube *V* and over the sponge *e′*. In the test tube receiving ordinary air the seeds germinate very well on the sponge *e,* while in the test tube receiving etherized air germination is suspended in the seeds resting on the sponge *e′*. In this way it has been possible to suspend germination for five to six days for garden-cress, which germinates within twenty-four hours, but once the test tube of ether *t* is removed and

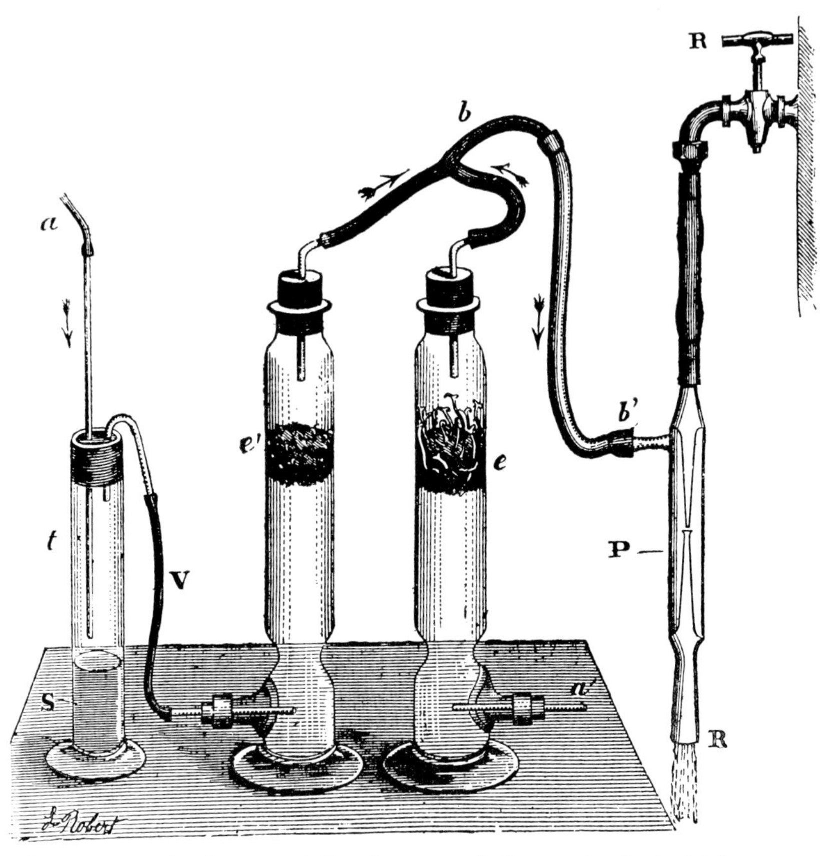

Figure 21. *a, a',* tubes permitting the external air to enter the test tubes.
b, b', branched rubber tube, carrying the air out of the test tubes and fitted to the air pump at its extremity *b'*.
e, e', moist sponges on which the seeds of garden cress are placed; on sponge *e* they have sprouted and grown.
t, test tube containing ether S at the bottom.
S, ether.
V, rubber tube carrying etherified air into the test tube with sponge *e'*.
RR, current of water flowing through the pump and producing aspiration through the apparatus.

ordinary air substituted for etherized air, germination was able to begin and progress rapidly.

I have repeated this experiment on quite a number of seeds;

on cabbage, turnip, flax, and barley, and always with the same results. The slowness of the germination is often an inconvenience, however. This is why I chose for the experiments in the lecture the seeds of garden-cress which are the most convenient of all because of their rapid germination.

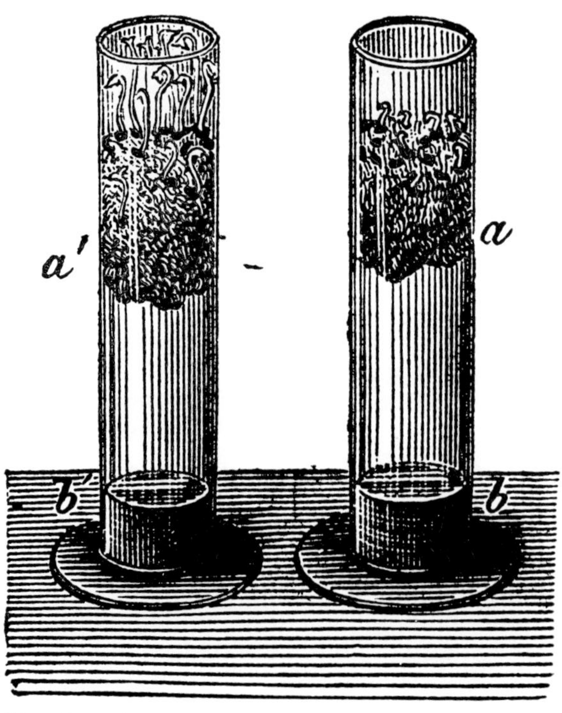

Figure 22.—Two footed test tubes in which the experiment on anesthesia of germination is arranged.
a, moist sponge on the surface of which are the cress seeds; *b,* chloroform water at the bottom of the test tube: the seeds have not germinated; *a',* moist sponge on the surface of which are the cress seeds; *b',* layer of ordinary water at the bottom of the test tube: the seeds have germinated.

These experiments on anesthesia of germination can be carried out by even simpler methods (see Fig. 22). It suffices, for example, to moisten the sponges *a* and *a'* on which the seeds are placed, the one *a* with ether or chloroform water, and the other *a'* with ordinary water; at the bottom of each test tube equal layers of etherized fluid in *b* and nonetherized fluid in *b′* are placed. Sometimes this arrangement fails, either because, on account of the ambient temperature, evaporation is not active enough and the sponge stays overloaded with the anesthetic agent and kills the seeds, or because on the contrary evaporation is too active and the anesthetic agent disappears and germination is not prevented, but only retarded.

I wanted to stabilize the experiment and make it very exact and as simple as possible to repeat. Here is how one ought to proceed: an ordinary footed test tube of about 130 cc is taken and into this test tube is introduced a small wet sponge sprinkled with seeds of garden cress, suspended in the atmosphere of the test tube by a thread. At the bottom of the test tube about 20 centimeters of distilled water are placed and the test tube stoppered. The next day, in the warm temperature of the summer, the cress seeds are in full germination. When now, in another test tube arranged exactly like the first, 10 centimeters of etherized water are added to 20 centimeters of pure water, and the test tube is stoppered as before, germination does not take place and remains in abeyance for four, five, six, or seven days; if the test tube is then unstoppered and the etherized water removed, germination reappears on the next day in the seeds in which it had been suspended by the anesthesia.

We shall add only a detail relative to the preparation of chloroformed or etherized water. To prepare chloroformed or etherized water, two flasks are taken, chloroform is poured in one, and ether in the other, distilled water is added, and the flasks are stoppered and shaken. The excess ether rises to the surface of the water, and the excess of chloroform falls to the bottom of the flask, but in both cases the water is saturated with the anesthetic agent. This is the water used in performing these experiments.

We have said that the anesthetics distinguish the vital phenomena of *organization* from the purely chemical phenomena of *destruction*. Etherization of germination provides us a striking example of it. In germination, in fact, two kinds of phenomena take place: (1) phenomena of organic creation properly speaking, by virtue of which the seed germinates, sprouts, and develops its rootlet, its stalk, etc. and (2) associated chemical phenomena,

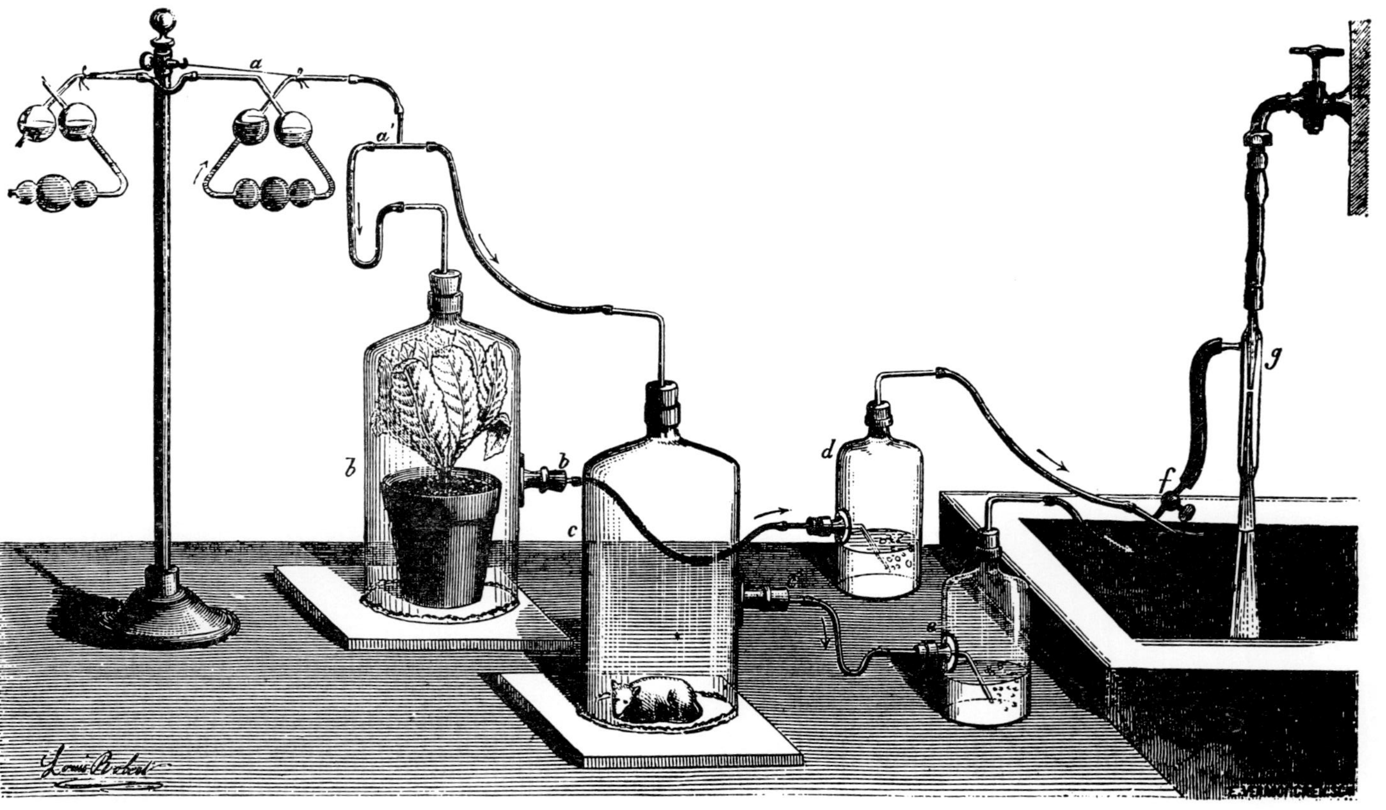

Figure 23.—Respiration of plants and animals.

which are, for example, the transformation of starch into sugar under the influence of diastase, the absorption of oxygen with the liberation of carbonic acid. But, in seeds whose vital phenomena of germination have been suspended by anesthesia, the chemical phenomena of germination are seen as usual; it can be ascertained that starch changes into sugar under the influence of diastase, that the atmosphere surrounding the seed is laden with carbonic acid etc.

It is thus demonstrated that the anesthetized seed whose growth is arrested, respires like the normal germinating seed. For this is suffices to place some baryta water at the bottom of the stoppered test tubes; in each case an apparently equal quantity of barium carbonate is precipitated.

We consider the respiration of living beings to be identical in both kingdoms and akin to a phenomenon of destruction characterized by the absorption of oxygen and the liberation of carbonic acid in plants as well as animals. This is true not only for the germinating seed, but also for the adult plant. In the latter, however, the respiratory function is more or less masked by the action of chlorophyll.

For a long time we have demonstrated in our courses this identity of respiration in animals and in plants by means of the apparatus above (see Fig. 23).

In the laboratory, in diffuse light, a young cabbage is placed under the bell-jar *b;* a white rat is placed under another bell-jar *c*. The cabbage and the rat both respire, as will be seen. A current of air is passed into the two bell-jars by means of a pump *g* that aspirates the air. A tap *f* makes it possible to moderate or accelerate the current of gas. The air entering the apparatus is deprived of the least traces of carbonic acid by its passage across two Liebig tubes with baryta water; the second tube serves as a control, its contents should remain perfectly clear. The current of air in *a′* is divided into two parts, one which traverses the bell-jar with cabbage *b,* and leaves at *b′* to pass to the flask *d* and pass through the baryta water which is very manifestly beclouded by the formation of barium carbonate; the other part of the current of air goes into the bell-jar of the rat *c,* and leaves it at *c′* to enter a similar flask of baryta where a cloud and a deposit of barium carbonate is equally seen to form.

It can be certain that the earth of the pot in which this cabbage is planted cannot be a source of error in this experiment.

Thus the plant respires like the animal, and the supposed antagonism in the respiration of animals and plants does not really exist.

Anesthesia of Eggs

I have tried at various times to anesthetize chicken eggs, housefly eggs, and silkworm eggs, working under appropriate conditions and using the air current apparatus described above (see Figs. 11 and 23). I have never succeeded in it. The eggs have developed very well in the test tubes receiving ordinary air but in the other they have been killed, that is to say, the arrested development was not resumed when a current of ordinary air was substituted for the current of etherized or chloroformed air.

I would not attempt to say that it would be impossible to succeed by adopting better conditions. I call attention to these trials only to show that the life of the seed and the life of the eggs are not comparable, as I have already remarked elsewhere with regard to latent life. Anyway, I repeat, one might succeed by studying more closely the conditions that ought to be employed. Henneguy, under the direction of Balbiani, has carried out and published some interesting observations on the action of anesthetic and other substances on the eggs and spermatozoa of fish.

Anesthesia of Organized Ferments

My experiments have been concerned especially with brewer's yeast. I have pursued them quite far. Today I shall limit myself to a simple reference, reserving for later a return to the details of this important subject.

One of those little tubes that we usually use for the study of fermentation is taken, and chloroformed or etherized sugar solution is placed in it, then brewer's yeast is added. In another similar tube brewer's yeast is added to ordinary sugar solution. The two tubes are left at a low temperature for twenty-four hours, for the anesthetic agent to have time to act on the yeast cells. The two tubes are placed in a waterbath at 35°C and soon the formation of gas is seen to develop actively in the tube containing ordinary sugar solution, while it does not take place in the other tube. But if the contents of this tube are thrown out on a filter so as to wash the yeast by a current of water for a sufficient time, and the yeast is replaced in ordinary sugar solution, the fermentation is seen to resume after a certain time. Müntz had already pointed out the influence of pure chloroform in arresting the fermentation of yeast. Bert has observed a similar influence of compressed air; in this case there was no anesthesia but destruction of the yeast, while in our experiments it was a matter of true anesthesia, since the yeast regains its properties as a ferment which the ether had caused to disappear momentarily.

Studying the anesthetized cells of yeast under the microscope, modifications produced in the protoplasmic contents of these cells are recognized, which explain the observed effects to us.

The Nonanesthesia of the Soluble Ferments

An interesting fact is the impossibility of suppressing the activity of the soluble ferment by anesthetics.

We shall limit ourselves here to a simple mention, not wishing to anticipate the studies that we are pursuing at this moment in view of our next course *on the fermentations.*

If diastatic ferments from animals or plants are dissolved in water containing chloroform or ether, it can be ascertained that their activity is not at all altered or diminished, on the contrary it seems to be more energetic, up to a certain point. It is the same for the inversive ferment of animals or plants. This explains to us why when yeast is placed in etherized sugar solution with sucrose, the phenomena of alcoholic fermentation do not appear, while those of the inversive fermentation of sucrose into glucose operate perfectly.

According to this, the fermentations might be divided into two groups; fermentations by protoplasmic or living ferments, which are arrested by anesthetics and nonprotoplasmic fermentation, produced by agents that are not endowed with life and which cannot be anesthetized.

It is in this way that chloroform and ether become, as I have said elsewhere, true vital reagents.

Anesthesia of the Activity of Chlorophyll in Plants

I have studied the action of anesthetics on the aquatic plants Potamogeton and Spirogyra. This is how I set up the experiment.

Under a bell-jar with a tube on top, filled with water containing carbonic acid, I place aquatic plants of the kinds mentioned, then, the entire jar being immersed in a large vessel, I cap the tube of the jar with a test tube, also filled with water, and destined to collect the gases given off by the plants. I place two jars so arranged in the sun, only in one of them I have placed along with the plants a wet sponge imbibed with a little chloroform. In the first jar, without the chloroform, a nearly pure oxygen is given off in good measure; in the second jar, with chloroform, only a very little gas is given off, which is carbonic acid. When after a sufficient time for the test to demonstrate that the chlorophyll of the plant has become unable to give off oxygen, I take out the plant and wash it well in running water and replace it in the sun without chloroform, I see the faculty of giving off oxygen in the sun reappear, which had been temporarily suspended.

We ought to remark on an interesting fact among those that we have just noted, namely, that the anesthetized aquatic plant gave off carbonic acid. This fact is in accord with what we have seen above; that the chemical phenomena of vital synthesis are alone abolished by anesthetics, while the chemical phenomena of destruction are not. In fact, the formation of carbonic acid by the act of respiration is not a vital act, since as Spallanzani showed, muscles separated from the body, inert and deprived of life, still form carbonic acid. A slice of cooked ham placed under a bell-jar respires and produces carbonic acid.

Thus with the help of anesthesia one can separate the activity of chlorophyll in plants, which is protoplasmic or vital, from respiration, which, like that of animals, is of a purely chemical nature.

Anesthesia of the Worms of Rusty Wheat

I have made but a few experiments on the anesthesia of lower animals.

Ether or chloroform kill infusoria very rapidly. I was not able to succeed in regulating their action. It is not the same for the worms in rusty wheat, which survive this kind of experiment very well.

We have seen, with regard to latent life, that dessicated worms from rusty wheat have the property of reviving when they are immersed in ordinary water. They do not manifest this property when they are immersed in water containing chloroform or ether, although it is generally necessary to dilute the etherized or chloroformed water by adding as much or more ordinary water, without which this worm will definitely be killed. In anesthetic water, sufficiently diluted, the worm remains immobile and does not come back to life; but it revives as soon as it is taken out and placed in ordinary water.

Examining under the microscope worms submerged in anesthetic immobility, several modifications in the appearance of their body can be ascertained. It appears to be more grainy, as though there had been a slight coagulation of its substance.

The facts that we have cited above and which we could multiply still further, demonstrate that anesthetic agents suspend the irritability of all living parts by acting in a physical way on their protoplasm, which is considered to be the seat of the irritability. From this we can understand with ease how the vital function is suspended, when the *irritability,* which is its *primum movens,* is suppressed.

If now we were to summarize in a general conclusion all our experiments made on man, on the higher animals, on the lower animals, on plants, on seeds, on eggs, etc., we would finally say that the anesthetics act both on *irritability* and on *sensibility.* What does this mean? Are irritability and sensibility thus identical, or if they are different, how is the common action exerted by the same agents to be understood? These are the important questions that must now be examined.

III. ON IRRITABILITY AND SENSIBILITY

Protoplasm enjoys the remarkable faculty of movement, of changing form under the influence of excitants; it is contractile. This faculty of movement is apparent in all naked protoplasmic masses, in the embryonic elements of connective tissue, the white blood corpuscles in higher animals, and in the ameba and myxomycetes among the lower beings.

Motility and *irritability* are moreover two interrelated properties that cannot be separated one from the other; movement is in fact produced by the influence of an agent; the agent is the *excitant,* the faculty of reacting to the excitant by a physical, mechanical, or chemical manifestation, is *irritability.*

We maintain that in irritability an elementary form of sensibility must be recognized, and in sensibility a very superior form of irritability, that is to say the property common to all tissues and to all elements of reacting, according to their nature, to external stimuli.

As we have often repeated, Linneus placed the criterion of animality in sensibility: *Vegetalia vivunt, animalia sentiunt",* he said.

To the celebrated naturalist of Uppsala, sensibility was the characteristic attribute of animals; and following him, his successors saw in the existence of this property the way to distinguish the two kingdoms of living nature, the proof of its duality.

Examining what in last analysis is this sensibility which has been made to constitute the superior mode of animal life, we do not encounter a simple *property,* but a complex vital manifestation corresponding to a *function.*

A distinction must be made between the functions of a living being and the *properties* of the organized substance which support them. Sensibility would be a complex phenomenon peculiar to certain beings, but which is related nevertheless to a simpler general phenomenon, irritability. Broussais, as we have said, expressed this opinion in part by accepting only one essential property of organized substance, *irritability,* involving as a consequence sensibility, contractility, and all the other secondary phenomena. As we have already seen, Virchow professed the same opinion; according to him, the vital phenomena have irritability as their immediate condition, a generic term including all the other vital properties.

It can be said that this doctrine existed in an embryonic state in Bichat.

The word alone is not clear; Bichat in fact retained the word sensibility, the source of so much confusion, but it is easy to see that he meant it in the sense that we now understand irritability, which at his time was not yet clearly distinguished. In animals he recognized an *animal sensibility,* and on the other hand a *vegetative* or *unconscious sensibility* residing in the organs of vegetative life, and manifesting itself by the visible acts these organs carry out when they are provoked by an external stimulation. But it may happen that this reaction to stimulants, artificial or physiological, is not betrayed by any movement or by any visible sign, and that it nevertheless exists, that it is linked to nutritive activity which is manifest only by its effects; this is what happens in plants, and Bichat accorded to plants and to certain parts of animals an *insensible sensibility,* that is to say, giving no sign of its existence.

Whatever one might think of these designations, *conscious* sensibility, *unconscious* sensibility, and *insensible* sensibility, one is nevertheless forced to recognize that they represent facts, and that they correspond to an exact impression of reality. All the acts of the organism are acts provoked by internal or external stimuli, physiological, normal, or artificial, they thus require a *sensibility* if one does not see in this word more than the faculty of reacting to the excitant. Now it is certain that in this reaction all degrees are to be found from an invisible *purely nutritive or trophic reaction* to a *motor reaction* under the purview of the senses, and finally the *conscious* reaction.

Thus to physiologists the term sensibility would present a meaning totally different from that attributed to it by the philosophers. From this has arisen a perpetual misunderstanding between the one and the other.

Generally the philosophers give the name of sensibility to the *faculty that we have of experiencing agreeable or disagreeable psychic changes as the result of bodily changes.*

It is in this sense of *conscious reaction* that the word is employed in everyday language.

It is easy to understand that in speaking of sensibility, physiologists ought not to consider it from such a restricted point of view; they cannot consider it as reduced to psychic changes in the consciousness, in the *Ego,* which are the sole preoccupations of the philosophers. These psychic changes escape the physiologist, who studies, and knows only material and tangible facts, even when they are totally foreign to the *Ego*. Such manifestations of sensibility lose all existence and all meaning when animals are considered and when man leaves his *inner forum* and the domain of his consciousness.

For physiologists sensibility is not only a fact of consciousness, it is in addition accompanied by material and perceptible manifestations that can serve as the basis for a physiological definition.

The phenomena of sensibility are in reality complex acts which are assisted by numerous secondary factors.

In man, and at the highest level of complexity, sensibility constitutes a function of the nervous system, a function that exists to harmonize the lives of the cells by satisfying the need of each cell to be excited and influenced by cosmic or organic agents external to it.

In a word, the nervous system answers a need that the organic elements have to be influenced one by the other, just as the respiratory and circulatory apparatus respond to the need experienced by the anatomical elements to be influenced by oxygen, etc.

The phenomenon of sensibility includes the ensemble of the following secondary facts:

1. *Impression* of an external agent (mechanical action on a peripheral nerve)
2. *Transmission* of this impression as a purely material or mechanical disturbance to the nervous centers, where it is transformed
3. Psychic phenomenon of *perception* (which can be lacking)

The impression, the transmission, purely material disturbances of the nerve centers, produce a *physical* change, that is to say, one of the same nature, within the nerve centers. Physiologists have called this *crude sensation* or *unconscious sensation.* The phenomenon does not stop there; the disturbance, bringing into action parts connected one to another, continues and is reflected onto nerves of movement and usually evokes a motor reaction, (movement, cry), and sometimes reactions of another kind nutritive, trophic, secretory, that are more difficult to perceive (icterus, pallor produced by an emotion, etc.).

Thus the phenomenon of sensibility in man himself, taking the expression in its usual sense, instead of being a simple vital property, is thus a most complex manifestation. Already it can be seen that it includes two kinds of phenomena: (1) purely material phenomena, a motor reaction, etc., resulting from the impression of an external agent, and (2) psychic phenomena.

If then we put aside the psychic phenomenon, there remains for us, to characterize sensibility, an ensemble of organic phenomena having as their starting point the impression of an external agent, and having at its end point the production of a variable functional act, movement, secretion, etc.: that which characterizes sensibility is the *material reaction to a stimulus*.

When the material or motor reaction is absent, we lose all possibility of appreciating the phenomenon of sensibility in animals. Outside of ourselves, of our consciousness, we have no information other than the production of motor responses; when we see them occur in an animal, we affirm that sensibility is in operation, when they are absent, we can affirm nothing. Thus the most general element of sensibility, and consequently the most important for the physiologist, is the reaction that terminates the cycle of material events, and which is at times mechanical, at times physicochemical.

Often, in fact, it is not always the motor element that responds to the excitation. There is often a molecular reaction of a different kind than this movement of translocation, which rarely appears except in the more highly organized animals, and which is lacking in plants. In any event there is a molecular reaction in all cases.

Sensibility is reduced to the motor reaction in the case of the *reflexes* proper; *reflex sensibility, excitoreflex capacity,* where the motor reaction takes place by itself without the intervention of consciousness. Thus for physiologists, apart from *conscious sensi-*

bility, there is an *unconscious sensibility,* an expression that appears to the philosophers as a real abuse of words.

On the other hand, the motor reaction can be absent in an animal poisoned by curare; the sensitive process then stops with impression, transmission, and perception, *without the motor reaction.* No obvious phenomenon betrays it and it would escape the physiologist if he did not have recourse to some expedient. But even though no manifest reaction takes place, it would not be necessary to characterize sensibility by the psychic phenomenon of sensation, for there could be other reactions which though not apparent are none the less real. There are physiological and material events, such as the molecular disturbance in the nerves, the special activity of the cerebral cells, and although these events are by no means detectable by the usual means, it is enough that they exist, and that appropriate expedients reveal them, to permit us to say that the sensitive process has still taken place. We shall not report all the particular examples that we could cite. We shall limit ourselves to the general features of a subject which would require much progress before we could treat it fully.

In résumé, what is peculiar about sensibility is *the reaction to stimulation from external agents.* This reaction is ordinarily motor if the organs of movement are in a state to show it; it can be of still another nature; trophic, secretory, or otherwise. When one gets to the roots of the phenomenon of sensibility, one finds nothing more than this; the faculty of transmitting, while modifying, the stimulation produced at one point, in such a way as to provoke in each organic element the initiation of its own proper activity.

Having come to this point we easily appreciate the cause of the misunderstanding between the philosophers and the physiologists. To the former, *sensibility is the whole of the psychic* reactions provoked by external modifiers; to the latter, for us, *it is the ensemble of the physiological reactions of every kind, evoked by these modifiers.*

Since a reaction can be envisaged in the cell, the organ, or in the apparatus that responds to the excitants, *sensibility would be the ability to respond, either by the total organism, the whole nervous system, or by one of its parts, or by a single cell.*

The ability of the cell to respond is the irritability, the sensibility, of the cell; in the same way, the ability to react by the whole of the nervous apparatus, or *conscious sensibility* can be considered as the irritability of this apparatus as a whole, *Unconscious sensibility* is the reaction of a part of this apparatus, a secondary sensibility.

In the infinite variety of beings the nervous system can be lacking in one of its parts, or as a whole, and yet life does not exist in greater measure than in the simplest organism such as a monocellular organism. Sensibility, that *physiological basis of life,* cannot be absent for all that. Thus irritability, that sort of *simple sensibility,* exists in the protoplasm of the cell, it is the elementary, irreducible property, while the reactions of nervous apparatus or organs are no different, and are no more than manifestations of refinements.

Sensibility, in its old meaning, considered as a property of the nervous system, is thus only a higher stage of a more simple property that is found everywhere; there is nothing essential or specifically distinct about it; it is the special irritability of the nerve, just as the property of contraction is the special irritability of the muscle and the property of secretion is the special irritability of the glandular element. Thus those properties on which were founded the distinction between plants and animals do not touch their very life, but only those mechanisms by which that life is exercised. Basically all these mechanisms are subject to a general and common condition: *irritability.*

Experimentation confirms and establishes these views solidly.

In fact the experiments with anesthetics prove that the same agent first destroys and suppresses *conscious sensibility* then *unconscious sensibility,* and then *insensible sensibility* or *irritability.* These suppressions are differing degrees of the action of the same agent, and as a result the phenomena themselves are different degrees of the same elementary phenomenon. The identical way in which they are influenced by the same reagent proves their identity, which becomes completely evident when the simple and clear conditions of the experiment are kept uppermost in mind.

In résumé, from the physiological point of view we are necessarily led to accept the identity of sensibility and irritability* because of the identity of the action of anesthetics on these vital manifestations. For in experimental physical science we have no other way of judging, if it is not to consider as identical those phenomena that exhibit identical physical characteristics.

Properly speaking, the anesthetic agent does not therefore act upon sensibility; it always acts definitely on *irritability,* and never on anything else, despite appearances. The irritability of the protoplasm of the cerebral cells is affected by the ether, and in this way the conscious sensorial function is abolished. In the same way, the protoplasm of the cells of the spinal cord, or of the nerve ganglia being altered, the functions of unconscious sensibility would be abolished in the corresponding nervous mechanisms. In a word, sensibility would be a *function,* irritability would be a *property;* it is the sole property that we influence.

If we were to penetrate more deeply into the analysis of the phenomena that we are examining, we would see that in reality irritability, just as well as sensibility, or the sensibilities, as all the vital properties, as well as all the functions, are the creations of our mind, metaphysical representations, upon which we cannot, in consequence, direct our action.

We do not in reality act on irritability, which is something immaterial, but rather upon protoplasm, which is material. By their contact with the protoplasm of the nerves, ether or chloroform produce a physical action, little known as yet, but real. Thus we always act on the material and not on the vital properties or functions. In a word, at the basis of all phenomenal manifestations of whatever kind they might be, there are nothing but physical conditions. It is only this that is tangible. Only the interpretations that we give to these physical phenomena are always metaphysical because our mind cannot conceive things and express them in any other way.

Metaphysics is related to the very essence of our mind; we can only speak metaphysically. I am thus not one of those who believe that we can ever suppress metaphysics, I believe only that its role in our concepts of the phenomena of the external world ought to be carefully studied, so as not to become the dupe of the illusions it can create in our mind.

---

*See my conference at Clermont-Ferrand, *Revue Scientifique.* no. 7, 18 August, 1877.

EIGHTH LECTURE

ORGANIZED SYNTHESIS, MORPHOLOGY

SUMMARY: Protoplasm represents life alone without specific form—Form is necessary to characterize the living being—Morphology is distinct from the chemical constitution of the beings.—
I. General Morphology—Four processes: (1) Cellular multiplication; (2) Rejuvenation; (3) Conjugation; (4) Gemmation.
II. Special Morphology—Development of the primordial egg—Ovogenic period; theory of encapsulation of germs; epigenesis—Period of fertilization—Embryological period.
III. Origin and Cause of Morphology—Morphology derives from atavism, from the previous state—Distinction between morphological synthesis and chemical synthesis—Final causes; they are merged with primary causes, and have no separate existence.

As we have already said, it is important in the living being to make a distinction between *matter* and *form*.

The living material, *protoplasm,* has no morphology by itself, no complication of form, or at least (and it comes to the same thing) its structure and elements are identical. In this amorphous, or rather *monomorphous,* matter life resides, but it is a *non-determinate* life, which is to say that here are to be found all the essential properties of which the manifestations of the higher beings are only diversified and definite expressions, or higher modalities. In the protoplasm are to be found the conditions for chemical syntheses that assimilate ambient substances and create the organic products; here irritability is encountered, as we have shown, as the starting point and the special form of sensibility.

Thus the protoplasm has everything it needs to live; all the properties that manifest themselves in living beings pertain to this material. Nevertheless protoplasm alone is only living matter, it is not a *living being*. It lacks the form that characterizes *definitive life.*

In studying protoplasm, its nature, and its properties, life is studied, so to say, in the naked state, as life without special *form.* Plasma is a sort of vital chaos, which has not yet been given shape, and where everything is intermingled, the faculties of disorganizing and reorganizing by synthesis, of reacting and of moving, etc.

The living being is a *structured protoplasm;* it has a specific and characteristic form. It constitutes a living mechanism whose real agent is the protoplasm. The *form* of life is independent of the essential *agent* of life, the protoplasm, since this remains the same through infinite morphological changes.

Form is therefore not a consequence of the nature of the vital matter. An essentially identical protoplasm could not give rise to so many different shapes. The morphology of the animal or of the plant is not to be explained by a property of the protoplasm.

This is why we separate morphological synthesis, which creates the forms, from the organic synthesis that creates the substances and the amorphous living matter. It is as it were a new level of complication in the study of life. After having determined the conditions of the *ideal* living being, amorphous and reduced to substance, it is necessary to know the *real* living being, shaped and appearing with a mechanism, a specific form.

It is necessary to make two observations immediately, which have their own interest, the one relative to mineral and animal morphology, the other on the relation of form to substance.

Morphology is by no means peculiar to living beings, they are not alone in exhibiting specific and constant forms. Mineral substances are capable of crystallizing, these crystals themselves can associate to form diverse and very constant shapes; *clusters, asteroids, macles, prisms, etc.;* at other times the substances take forms that are not truly crystalline, glucose in hillocks, leucine in balls, lecithin in globes, etc.

There is therefore justification, up to a certain point, for relating the two kingdoms of mineral and living beings, in this sense that we see in the one and the other this morphological influence that gives a definite form to the parts. We know that the analogy does not stop at this primary general resemblance; the facts of regeneration of crystals pointed out before (see Lecture I) have shown us in the crystal something comparable to the tendency by which the animal repairs itself, and completes and reconstitutes the individual morphological type.

But the crystalline forms of minerals are, no more than the living forms, a rigorous, absolute consequence of the chemical nature of matter. Dimorphous substances are a very clear example; sulfur can exist in two incompatible crystalline forms and in the amorphous state; phosphorus and arsenious acid also show us a single matter shaped in different forms. The isomeric and polymeric substances of organic chemistry provide us still another proof of a different kind, that the identity of the substrate is compatible with a variety of forms, groupings, and phenomenal manifestations.

In other terms, there are, in mineral and inorganic chemistry, bodies of the same form that have different chemical compositions and bodies of different composition that have an identical form.

The study of forms no longer belong to chemistry and is not in any way explained by its laws. Chemistry is concerned with the composition of bodies; where morphology (that is to say the study of form) begins, chemistry, properly speaking, ceases.

The materials that the organism produces or puts to work are thus not only chemically constituted; they are also worked over morphologically and arranged in a more or less characteristic shape. It is even possible for the form to appear to be more essential than the material. This is so for the bony skeleton and the eggshell in birds. By changing the feeding of these animals and substituting salts of magnesium for calcium salts, it has been reported that the usual composition of the bones and the composition of the shell were changed and a certain proportion of magnesium had taken the place of the lime. I have often heard it said by the naturalist A. Moquin-Tandon that the same species of snails living on calcareous or silicious soil sometimes have silicon and sometimes calcium carbonate in the composition of their shells, without, indeed, their specific morphology being otherwise modified. These diverse substances would be interchangeable in all proportions in the formation of organs, and they would behave like isomorphic substances in the formation of crystals.

These comparisons between mineral forms and living forms certainly constitute only very distant analogies, and it would be imprudent to exaggerate them. It suffices to mention them. They ought simply to give us a better conception of the theoretical separation of these two types of vital creation: creation or *chemical* synthesis, and creation or *morphological* synthesis, which in fact are intermixed because of their simultaneity, but are nonetheless essentially distinct in their nature.

We must now study this morphological synthesis, first as to its products and then as to its causes.

The independence of form and matter is pushed still further in living beings than in minerals. Morphology, as we shall see, seems to be governed by laws that are absolutely different from those that govern the essential vital manifestations of the protoplasm. It is based upon this material and its properties, but it utilizes it in a completely independent fashion, and according to conditions that are not necessarily contained within it.

The varied forms that result from these morphological laws give rise to vital *phenomena,* very different from one another and which are nothing more than the expression of the *morphology of the being.*

The protoplasmic matter, as we have said before (see Lecture V), can at first constitute beings in some way without fixed form, or at least without morphologically distinct vital mechanisms. These are the simplest beings, possessing only naked life without the infinitely varied and diversified forms in which it will appear to us later on. These beings are in reality protoplasmic beings or cytodes of which Haeckel has made a group, even a kingdom, under the name *Monera.*

Among these moneran or protoplasmic beings there are first of all the amoebas. We represent here a fresh water moneran, the *Protamoeba primitiva* (see Fig. 24) and amoebas in their various changing forms (see Fig. 25). We point out that these

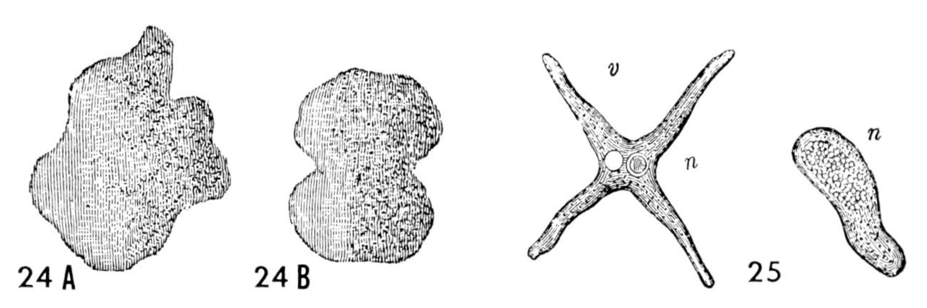

Figure 24.—*Protamoeba primitiva,* Haeckel
A, an entire Moneran.
B, the same Moneran divided in two halves by a median furrow.
Figure 25.—Two different forms of the Amoeba from mud.
n, nucleus
v, contractile vacuole.

amoeboid beings, which can live in the free state in the cosmic environment, can also live as an element of some sort within the internal environment of other higher beings. Thus in Figure 26 we see free living amoebas and amoebas of the blood or lymphatic corpuscles of *Lumbricus agricola* behaving in exactly the same way. Balbiani, to whose kindness I owe this figure, has seen that the amoebas of lumbricus can engulf little bodies in suspension in the blood, absolutely like the amoebas in infusions, which

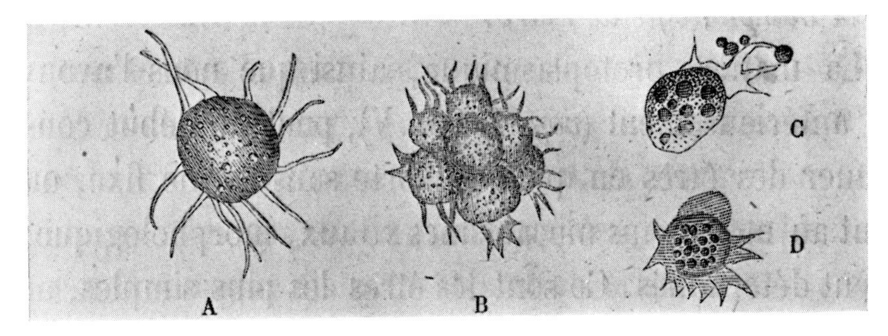

Figure 26.—Lymphatic corpuscles of Lumbricus and Amoeba from infusions.
A, an isolated lymphatic corpuscle from Lumbricus.
B, aggregated lymphatic corpuscles from Lumbricus.
C, Amoeba from infusions engulfing colored bodies.
D, Lymphatic corpuscles of Lumbricus having engulfed the same colored bodies (Prussian blue). (See the plate at the end of the volume).

proves indeed that they are the same beings. We reproduce also the figure of *Protogenes primordialis* discovered in 1864 by Haeckel (see Fig. 27 and Lecture V, page 137). It is necessary to point

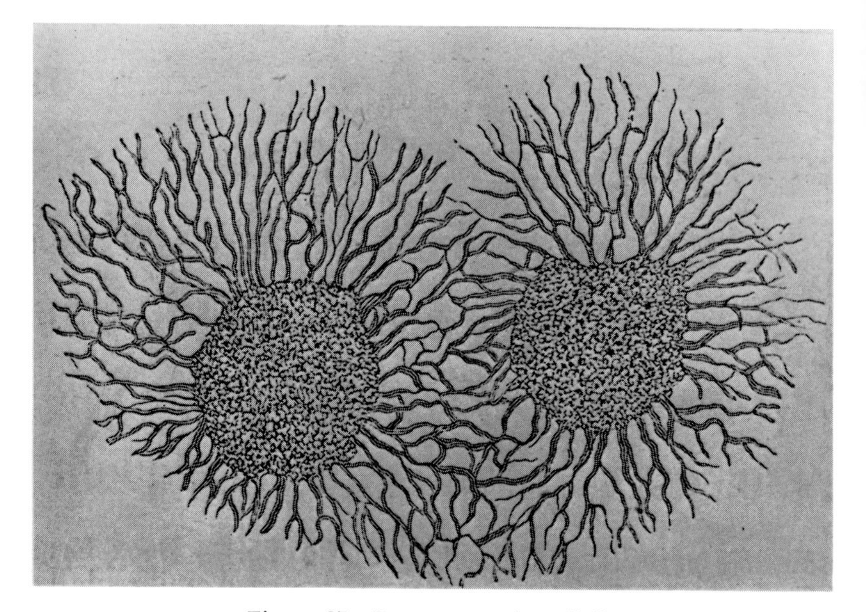

Figure 27.—*Protogenes primordialis.*

out, among these rudimentary beings *Bathybius haeckelii,* discovered in 1868 by Huxley, a species of gigantic amoeboid network which lives in the depths of the seas (Fig. 28 A and B, and Lecture V, page 136).

We shall not discuss the question of ascertaining whether the moneran beings have a true morphology; or whether the *cytode* of Haeckel can exist at one and the same time, by a sort of arrested development, either as an individual complete living animal or as the possible beginning of other much more complex organisms. These questions are most uncertain and most problematical. For ourselves, we accept a genuine morphology only when we see the same organic element start from a fixed point and follow regularly a developmental progression which brings it to an organic type equally fixed and determined in advance. But this development really begins with the cell.

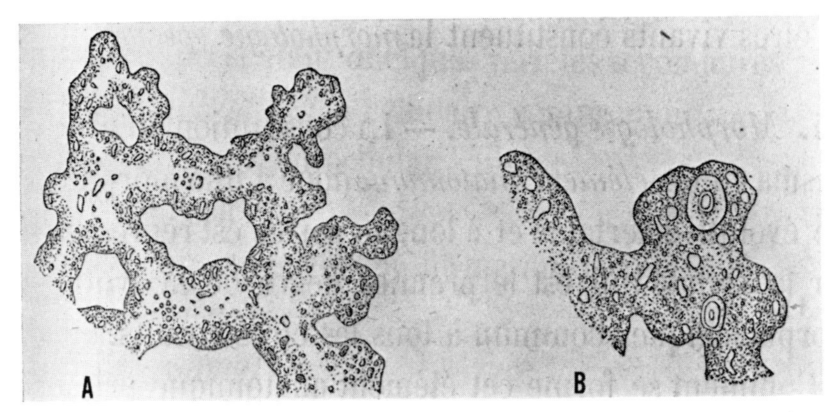

Figure 28A.—*Bathybius Haeckelii,* protoplasmic organism living at the bottom of the sea. The figure represents a small portion of the network of naked protoplasm.

Figure 28B.—Protoplasmic network with discoliths and cyatholiths found in other Monera which are apparently excretory products (Haeckel).

The cells take form, multiply, and accumulate to constitute at first the mass of the organism, then they are modified and give rise to specific forms that characterize from the start the beings that are to arise from them.

The mechanism of the formation and the multiplication of cells is what we call *general morphology*. The grouping of these cells and the specific configuration according to which they are arranged to form living beings constitute *special morphology*.

I. GENERAL MORPHOLOGY

The formation of protoplasm into an *anatomical element* endowed with a definite and long range developmental morphology is represented by the cell, which is the first stage of morphological synthesis common to all living beings.

How is this primordial anatomical element, the cell, formed?

We know that life exists, before the cell, in the protoplasm; but in the actual state of things we never see a cell appear developmentally without a previous cell. The axiom *"Omnis cellula e cellula"* remains true, therefore, for the two kingdoms. The histologists who have best studied the question have arrived at this conclusion: "The formation of cells in the absence of others in organic fluids or *blastemas,*" said Strasburger (1876), "is a hypothesis that has never been proven. Their spontaneous generation is no more correct than that of individual organic forms."

It is the opinion of the botanists as well as the zoologists that cells all arise from the protoplasm of a preexisting cell. "Every new formation of cells" says Sachs, "is basically only a new arrangement of preexisting protoplasm."

It is necessary to examine the processes by which the cell appears at the expense of a preexisting cell.

The processes of the genesis of cells are the same in the two kingdoms, as might have been expected.

Four principal forms of cellular genesis can be distinguished, exhibiting several secondary varieties:

1. Cellular *multiplication,* including:
   (a) the free formation of cells
   (b) division
2. *Rejuvenation* or complete formation
3. *Conjugation*
4. *Gemmation*

A. *Multiplication*

This is the process of cellular genesis in which there is a production of two or more elements at the expense of a single one.

It can happen that only a portion of the protoplasm of the original element participates in the formation of the new elements. This then is what is called *free cellular formation.*

Plants and animals provide examples of this. This is how the endospermic cells of Phanerogams are formed in the interior of the embryonic sac, and at the expense of a portion only of the protoplasm contained therein (see Fig. 29).

In animals Balbiani has observed this type of genesis in the constitution of the blastodermic cells of insects at the expense of the vitellus. One part only of the vitellus provides the new cells (see Fig. 30).

If all the protoplasm of the original element is employed in the constitution of new cells, this then becomes the process of *division.*

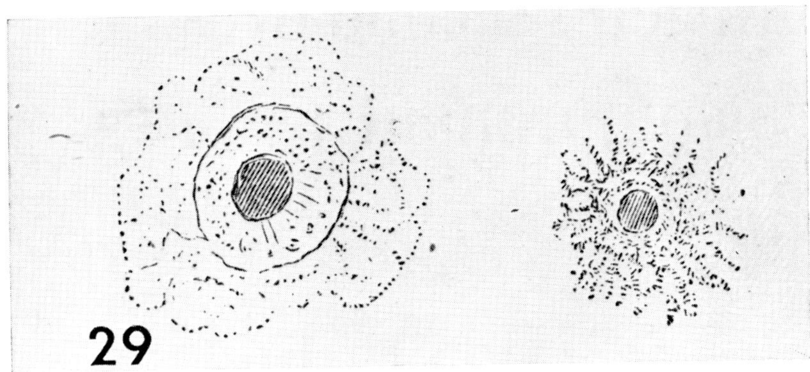

Figure 29.—Free formation of cells in the endosperm of *Phaseolus multiflorus,* first form. (Strasburger, p. 501)

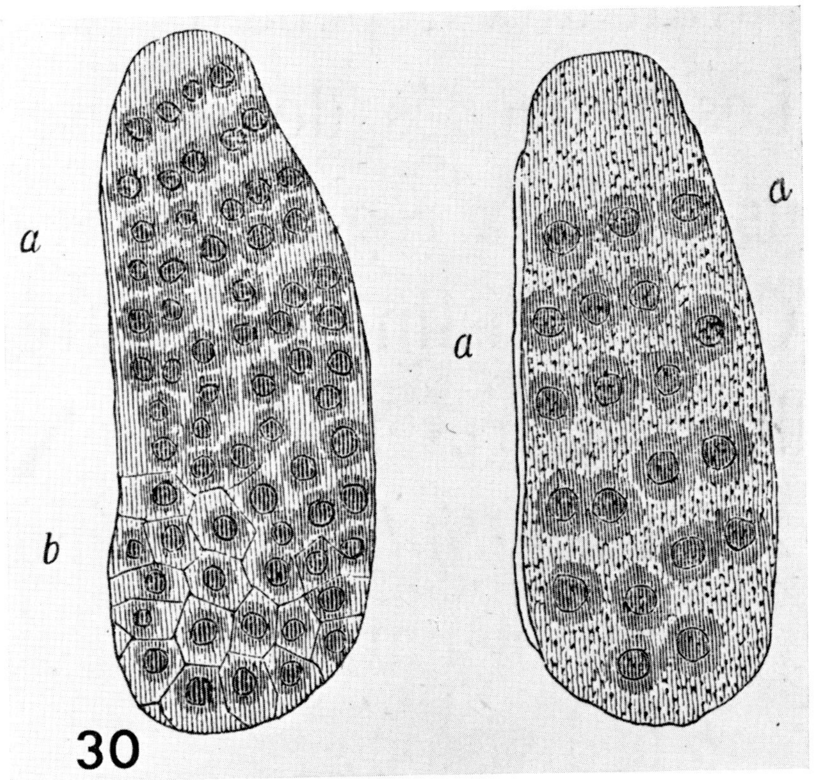

Figure 30—Genesis of cells by free formation in the blastodermic layer of an insect egg. (Balbiani).
a, formation of nuclei.
b, differentiation of cells.

This process of division is the most general of all. The greatest number of the plant elements is produced in this way. As to the elements of animals, it has been accepted for a number of years that *division* is their sole origin. This kind of genesis, which Remak reported in 1850, in a study of the cells of the blastoderm, has been considered to be the exclusive means of cellular genesis. This is the opinion of Kölliker.

Division is thus the most universal genetic method. A cell divides and produces two new ones. There can be two possibilities; either the primitive envelope has no thick wall at all, or there is a well developed envelope. In the first case there is a *simple scission,* in the second, *endogenous division.*

The monera, the amoebae, the infusoria, and the blood corpuscles of the embryo divide in this way. The mass of protoplasm making up these animals prolongates, develops a constriction, and soon separates into two new masses; each one constitutes from now on a distinct individual, in which the same process of vital phenomena begins again (see Fig. 24).

As to *endogenous division,* it was described a few years ago in the simplest manner. The nucleus, it was said, takes the initiative and within the nucleus, the nucleolus. In place of a single nucleolus, two of them are perceived, then the nucleus constricts and segments, carrying along the new nucleolus. The division of the nucleus involves that of the protoplasm, and finally instead of one cell there are two of them.

But this idea, that was maintained up to these last few years, was not the real expression of the truth. We have already mentioned new investigations that tend to modify these overly simple views (see Lecture V, page 141). We must return to them.

Strasburger has studied the productions of cells at the organic crest of the embryonic sac in several plants, and in particular in the conifers, *Picea vulgaris* (*see* Figs. 31, 32, 33, 34, 35).

At first, the protoplasm of this sac gives rise from one of its parts to four cells arising by *free formation.* These are cells that later on lend themselves well to the study of division and the circumstances that accompany it.

Two successive phases can be distinguished. In the first phase the nucleus of the protoplasmic mass exhibits two masses of gran-

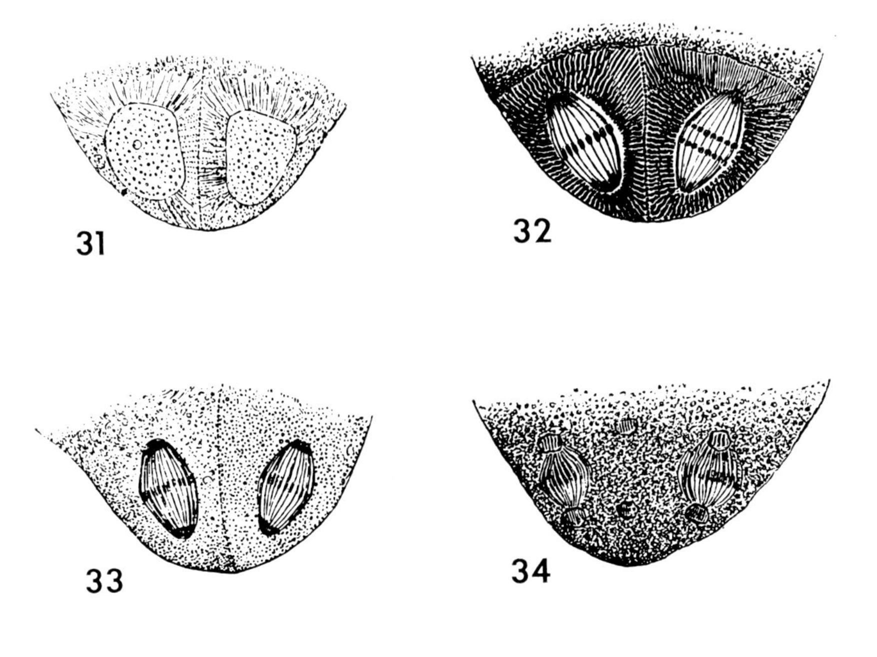

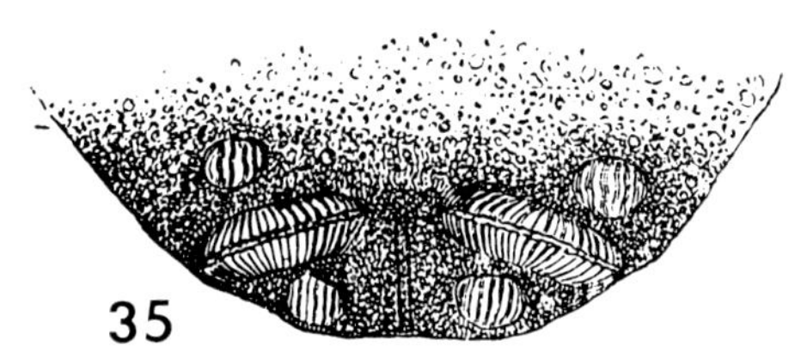

Figure 31, 32, 33, 34, 35.—Genesis of cells by division in plants.

Figure 31.—Nuclei appearing simultaneously in the egg of *Pinus sylvestris* (Strasburger, p. 250).

Figure 32.—Beginning of nuclear division in the egg of *Pinus sylvestris*. The nucleus at the right shows a more advanced stage than at the left (Strasburger, p. 260).

Figure 33.—More advanced state than in Figure 27 (sic). The cellular plaques are already conspicuous at the equator between the new nuclei in course of formation (Strasburger, p. 250).

Figure 34.—The formation of new nuclei has been completed; the cellular plaques are more marked (Strasburger, p. 250).

Figure 35.—The cellular membrane already secreted in the middle of the cellular plaque (Strasburger, p. 250).

ules situated at the two poles, or opposite points; these masses are connected by intermediary filaments. These filaments, swollen uniformly at their center, constitute as a whole an equatorial disk or *nuclear disk*. This is what is seen on the left hand side of Figure 32. Then the swellings divide and return, each toward the corresponding pole. This separation and this movement are seen on the right side of Figure 32.

In the second phase, there forms along the equatorial plane a new series of swellings which as a whole constitute the *cellular plaque;* this cleaves in two; between the two cleavages a partition of cellulose is formed, and the process continuing, there are soon, instead of the primitive mass, two complete cells within the embryonic sac.

The nucleus does not always play this essential role in the genesis of the cell. Cases are known where it does not yet exist at the moment when the protoplasm divides, and cases where the nucleus does exist, but remains, so to speak, a stranger to the appearance of attractive centers, which will group the protoplasmic material to form two new cells from it.

These are the complex phenomena that have been observed in plants, and equally in animals, and which seem to have a very great general applicability. Bütschli (*see* Lecture V, page 140) has observed the division of embryonic blood cells in the chick (*see* Fig. 36) ; Weitzel, the proliferations of cells in the inflamed conjunctiva; Balbiani, the multiplication of cells in the ovarian epithelium of insects; Auerbach, Fol, Strasburger, and Klebs have encountered a considerable number of facts of the same kind. Interpreting these facts one is led to believe that in animals there exists but one unique process for the genesis of cells, to which the others are all related and are simply abbreviations of it.

These studies show us, in cellular genesis by division, something analogous to the operation of the forces of attraction and repulsion, acting especially on the nucleus, and manifested by the polarity and the radiating disposition they impress on the particles of the protoplasm.

B. Rejuvenation

Rejuvenation, or complete formation, is a rare process of which several examples are found in the plant kingdom; it is not

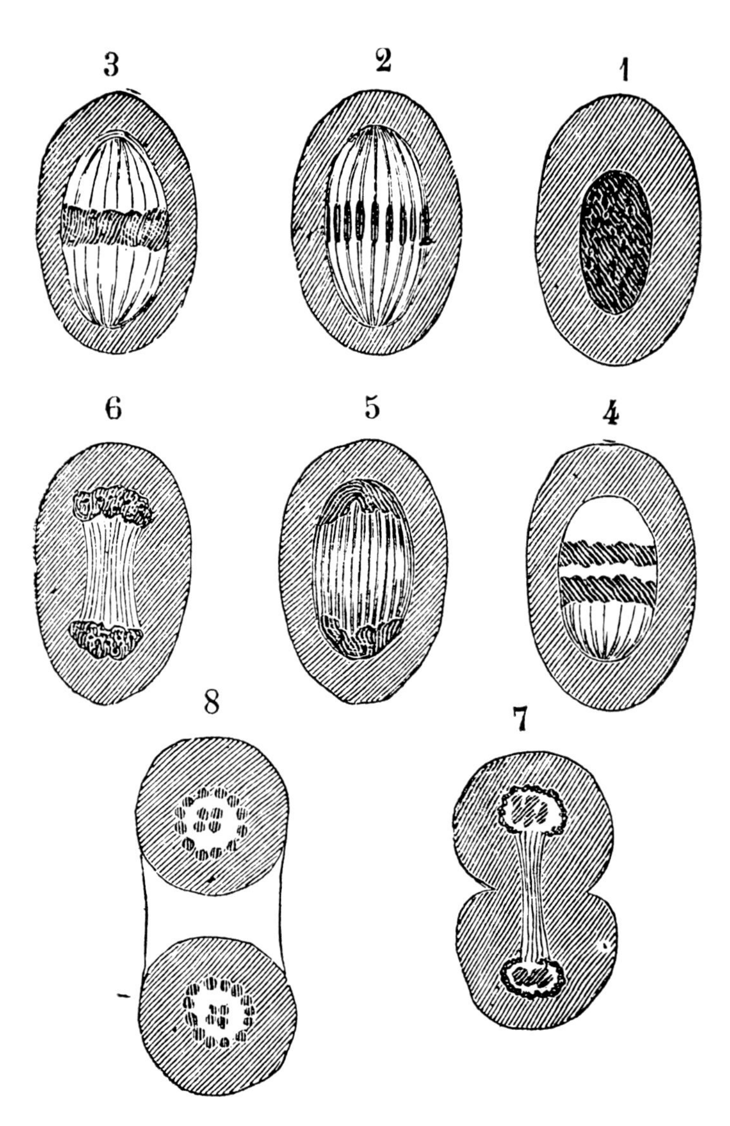

Figure 36.—Genesis of cells by division in animals. 1, 2, 3, 4, 5, 6, 7, 8, successive phases in the division of a blood corpuscle in a chick embryo, after Bütschli.

known at all in the animal kingdom. There is a preexisting cell: the entire mass of the protoplasm of this cell forms a new cell by a sort of renewal or simple rejuvenation of this protoplasm. It is by this means that Pringsheim saw zoospores form in algae of the genus *Oedogonium* (see Fig. 37).

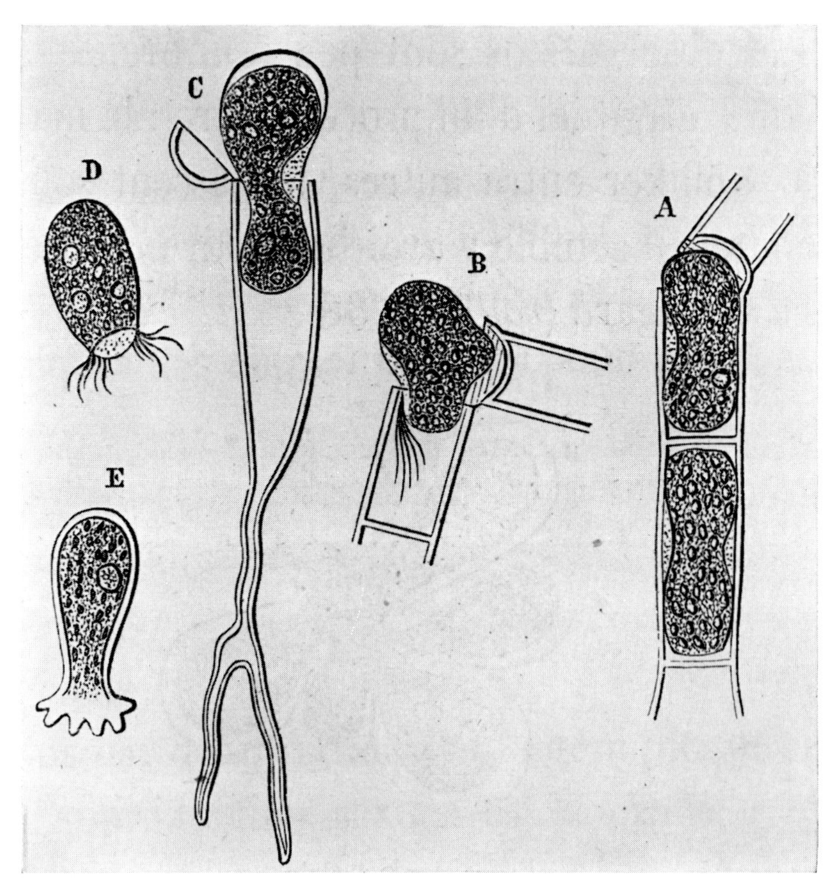

Figure 37.—Total formation by rejuvenation (Sachs, p. 12). A, B, exit of zoospores from an *Oedogonium;* C, exit of the whole of the protoplasm of a young plant of *Oedogonium* in the form of a zoospore; D, free zoospore in motion; E, the same after it has become fixed and has formed its adherent disk.

C. Conjugation

Conjugation consists in the fusion of two or more masses of protoplasm into a single one. Two elements participate in the formation of the new element, and this can take place in two ways, either by conjugation properly so called, or by sexual conjugation, that is to say by fertilization.

In ordinary conjugation the two cells that participate are identical in form and size. This is how the zygospores of conjugated and volvocinean algae, and the zygospores of the mold of myxomycetes and the mucorines are formed. The animal kingdom does not provide any known example of such cellular genesis (see the plate at the end of this volume).

As to sexual conjugation or fertilization, in which the two elements are differentiated, there are examples in the oospores of the cryptogams and in animals, a universal type in the fertilization of the egg.

D. Gemmation

Finally, we have called attention to a fourth kind of cellular genesis, *gemmation,* or budding. There are but a few observations, and it is certain that we are dealing here with a rare process; the majority of authors, Kölliker among others, pass over it in silence.

However, there seems to be a small number of positive facts about it (see Fig. 38).

Such, for example, are the formation of eggs by the budding of cells of the ovigeneous sheath in insects, the formation of polar globules observed by Robin; the multiplication of acinetic infusoria (Podophyra gemmipara) observed by Hertwig, and finally the division of the lymph corpuscles of the axolotl, which was observed by Ranvier. The nucleus elongates, constricts like a pouch, and then from this nucleus, more or less numerous buds are seen to arise, each of them having a nucleolus. Each of these buds seems to govern the surrounding mass of protoplasm, which it gathers around itself so as to form a new cell.

Such are the processes of *general morphology,* by which one cell arises from another cell, by which, in short, the simplest organism is constituted.

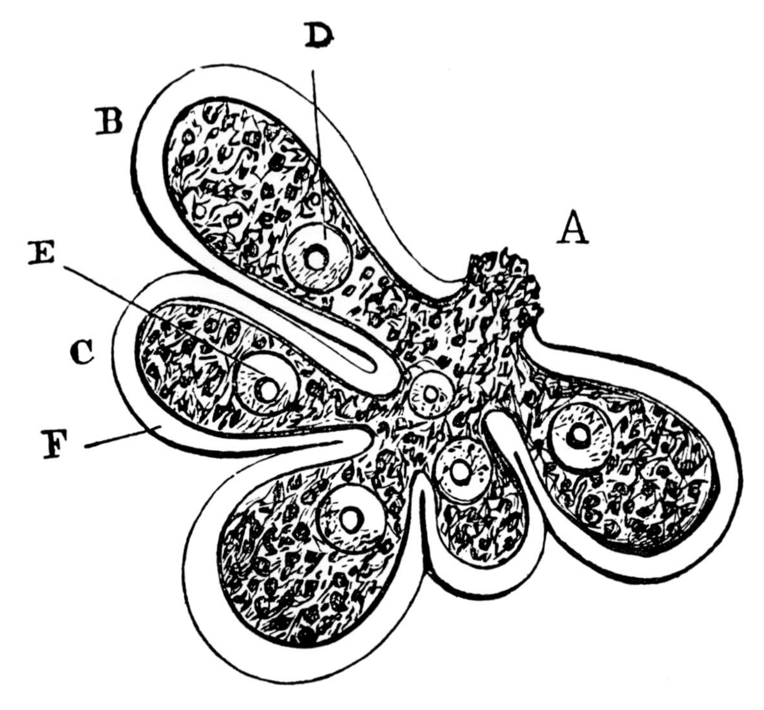

Figure 38.—Gemmation. Ovalation of a lamellibranch mollusk *(Venus decussata).* A, mother cell; B, C, buds formed by the retraction of the cell wall F under the influence of the new nuclei D, E, arising from the division of the original nucleus (after Leydig).

Now we shall examine *special morphology,* which presides over the production of the complex and specific forms of animals and plants.

II. SPECIAL MORPHOLOGY

The starting point of the animal or plant species is a cell called an egg or ovum.

To be sure, a certain number of beings proceed from parents by monogenic or asexual processes, but sexual reproduction is the genetic process par excellence, universal, and sufficient in itself to assure the perpetuation of the species.

The egg itself is originally a cell. Going back to its first appearance, it is found in all animals in the state of *protovum* or *primordial ovule,* it is formed from a mass of protoplasm or *primitive vitellus* or *archicyte, or primitive plasma,* a mass at the center of which there is a voluminous, granular, refringent nucleus, which is the *primitive nucleus,* or *vesicle of Purkinje.*

This primordial ovule so constituted is originally an epithelial cell, appearing within the maternal organism at the earliest stage of development; this cell becomes visible among the epithelial cells in the same layer, grows and soon characterizes itself as a primordial ovule.

The method of formation of this primordial ovule at the expense of a preexising epithelial cell and its constitution as a nucleated mass of protoplasm are absolutely general facts applicable to all animals from the protozoa to the vertebrates, as the embryological works published in the last ten years have established.

This is the common origin of all the living beings; this simple cell enjoys the faculty of giving birth, by a series of successive differentiations in the products of its proliferation, to the most complex specific forms.

The egg, in fact, does not remain indefinitely at the stage of the primordial ovule; it is an element essentially endowed with the faculty of development, which modifies, multiplies, completes, and differentiates itself by a progressive movement and continual labor. The individual animal at the completed state is, so to speak, only the most advanced stage or the ultimate phase of this development, while on the other hand the primitive ovule could be called the primary state of the animal, its beginning, or its first draft.

Pursuing his excellent studies on the organs of reproduction in the aphidians, Balbiani has been able to report still further on the origin of the egg. To him, the egg is not a simple anatomical element, it is already an *organism,* it is constituted by the union or conjugation of two elements, the one playing the role of the male element, the other the role of the female element; these two bodies whose union constitutes the ovule are on the one hand the *germinal vesicle* with its protoplasm and on the other hand the *embryogenic cell* or *androcyte.* This latter would not be a part of the maternal organism *already constituted,* but it would exist previously in the egg from which this maternal organism arose. There would thus be in the maternal egg an essential element of the egg of the offspring. This ovular element is transmitted; it persists, not only like an organ belonging to an individual who is its carrier, but like an element belonging to the ancestors, and which in the body of the present individual would constitute a veritable atavistic parasite. At first it was believed that the egg is a product of the maternal organism in the state of full development, then it was said that it was a product of the maternal organism from its embryonic state, and even before its sex was characterized. Balbiani makes one further step along this route of origin and relates the egg to the still undeveloped maternal organ, existing solely in potential, that is to say within the maternal egg.

The same can be said of the previous egg, and so on, going on back. Thus the egg contains an essential element of the eggs of successive generations; a specific and nonindividual element. This doctrine of Balbiani seems to a certain extent to revive the famous theory of the involution or of the *encapsulation of germs* proposed during the last century by the naturalist-philosopher Ch. Bonnet of Geneva. At the time the Genevan naturalist proposed his hypothesis it was thought that the new being existed preformed in the egg; others said in the seminal liquid; it was not the present being that created it, it only carried it so to speak, and provided shelter for this sketch or miniature of the offspring. Ch. Bonnet was led by his *a priori* meditations and by his experiments on plant lice to propose the *preformation* or *preexistence of the germ,* not only in the egg in which it will develop, but the indefinite preformation, for all time, of the egg itself.

The origin of this doctrine is found in the philosophic ideas of Leibnitz. Leibnitz considered all the phenomena in the universe as the simple consequence of a primordial act of creation. The creative power that had intervened a first time did not need to repeat its effort, and the natural order was fixed for the rest of time. In particular, the first being contained, in potential and in substance, all the generations that came after it, and the observer sees only the development of those aboriginal germs, included the one in the other.

It is this view that the Genevan philosopher Bonnet adopted. He admitted that an animal does not really create the beings whose source it becomes, that it simply contained the germs, enwrapped so to say, the one by the others and shedding their envelopes in succession. If certain testimony is to be believed, Cuvier, whose precise genius did not harmonize well with hypothesis, would have nevertheless received this one with favor.

The development of the science has discarded what in this doctrine was manifestly erroneous, namely that the egg is the reduced image of the new being which has only so to speak to deploy and grow. The animal is formed, not by amplification of already existing parts, but by the formation, the successive creation of new parts, or *epigenesis,* as we shall mention soon. As to the other part of the doctrine, which consists in supposing that the egg encloses, not only potentially, but in a formed and substantial form, some elements of the generations to follow is the part of the doctrine that the ideas of Balbiani have come to rescue from the neglect and the disfavor into which they had fallen.

In the history of the development, or of the ontogeny of an animal, three periods can be distinguished:

1. The *period of ovogenesis,* which lasts from the origin of the egg to its complete formation.
2. The *period of fertilization,* which corresponds to the moment when the egg, having arrived at the state of maturity, receives the new impetus resulting from the contact with the male element.
3. Finally, the *period of embryogenesis,* the longest, which includes the series of phenomena by which the fertilized egg is carried along to the complete development of the animal.

We need not detail here the history of these three periods; we need only to characterize them briefly, since they mark the three principal stages in morphogenesis.

We call attention to the common starting point of all organization in this universally identical form, the *primordial ovule,* a simple nucleated mass of protoplasm. This identical origin of all organized beings is an essential phenomenon indeed, and worthy of elucidating. It has been developed particularly since the work of Waldeyer in 1870.

This primordial ovule undergoes a development (ovogenic

development) that brings it to the state in which it can effectively be subjected to impregnation by the male element, that is to say, to the state of a ripe egg. This development includes three principal events: the formation of an envelope limiting the element on the outside, the *vitelline membrane;* the growth of the primitive protoplasmic mass by the addition of new elements constituting the *secondary vitellus,* or nutritive vitellus, or *paralecith* or *deutoplasm,* according to the various names that authors have given to it; finally, and in the third place, the nucleus, or *germinative vesicle* of Purkinje, until then homogeneous in all its parts now exhibits nuclear granulations, the *germinative spots* or Wagner's spots.

At this early period differences appear according to whether the egg is to form an animal of this or that zoological group. Before any fertilization, before any development, it is possible to predict from the particular anatomical characteristics of the ripe egg the general direction of its development and the group to which the animal it will form will belong. The vitelline membrane, for example, is radially striated in the mammals and the bony fishes and exhibits a micropile. Nothing similar is found in the birds. The secondary vitellus can have various proportions relative to the primary vitellus; sometimes it is very abundant, as is the case of oviparous animals, the birds and reptiles, and sometimes it is scanty, which is the case in the vivipara, such as the mammals. Finally the germinative spots in the nucleus are quite different in number in one or the other of the vertebrates; there are more than 100 to 200 in the fish, and on the contrary one or two in the mammals.

A study of ovogenesis extended to all the groups would thus have the result of showing a very early differentiation in the work of development. It seems indeed that from the common beginning the pathways diverge and each primordial ovule has the way fixed in advance along which it will progress without cease, until the completion, under the direction of the laws of morphogenesis, of the animal type that was in effect inscribed within it.

The second period in the development of the egg is characterized by the phenomenon of fertilization, and all the secondary events that prepare for it or are connected with it. As we have said, the egg is a most energetic plastic element, a center of chemical and morphological attraction. The developmental process of this element is reinforced in a still unknown manner by the intervention of the male element, that is to say, fertilization.

Once fertilization has been accomplished, developmental processes take on an intense activity and the embryological phase begins

The problem of embryogenesis consists, in the end, in explaining by what successive processes the single cell of the ovum gives rise to this polycellular construction of such complicated architecture that forms the living machine.

At first we had recourse to hypotheses, before appealing to observation, in the attempt to penetrate this mystery.

Two opposing theories present themselves to the mind of the philosophical naturalist, each of which has had its partisans; one is the theory of the *involution* of a part, the other the theory of *epigenesis*. Today the debate has been resolved, and it is known, from the works of the famous embryologist Caspar Frederick Wolff, that the organism develops from the egg by *epigenesis*.

The partisans of involution thought that the generation of a being was not a true creation. The offspring preexisted, completely formed, with its organs, its apparatus, its shape, in the germ, and fertilization did nothing but release it. This germ, a reduced image of the new being, was for certain naturalists the *egg*, from which derived their name of *ovists*, such as Swammerdamm, Malpighi and Haller.—For others, the *spermatists*, such as Leeuwenhoeck and Spallanzani, it was the *spermatic animal* which was the germ, but for the one and the other, the germ was the rough draft, the miniature of the embryo, and this was the essential point of the doctrine. The being did not therefore begin with the act of fertilization, it already preexisted, in a dormant state, awaiting only to be aroused from this lethargic state by the impetus of fertilization. Defended by Leibnitz among philosophers and by Haller among physiologists, this doctrine remained universally accepted until the moment when C. F. Wolff, the principal founder of modern embryology, gave it the mortal blow and revealed the true nature of organic development. "He proved that the development of each organism takes place by a series of new formations and that neither in the egg nor in the spermatozoan does there exist the least trace of the final forms of the organism." [Haeckel, *Anthropogénie,* p. 28 (1764)].

C. F. Wolff showed, in fact, by studying the development of the digestive tube in the chick, that there is a stage at which this apparatus is yet only a sort of ovular membrane, a *germinative layer,* that passes through a series of continuous transformations, and by new additions comes to form the intestinal canal, the glands connected with it, the liver, the lungs, etc. In this observation is found the germ of the discovery of the *embryonic layers,* which Baër completed and later introduced into the science.

Thus the parts of the body are made in succession, one after the other, by successive additions and differentiations. Nothing preexists in its final form and design. The human germ is not a *homunculus,* a reduced and perfect image of the adult, it is a cellular mass, which through slow travail acquires successively more complicated forms.

The primary phenomena by which embryogenic development begins are essentially the same from one end of the animal kingdom to the other. In mammals the protoplasmic mass that constitutes the fertilized egg segments into halves by an endogenous division. Each of the two new masses undergoes a similar segmentation. This phenomenon, called *vitelline division* ends, after these repeated divisions of the principal protoplasmic mass, in the formation of a mass of cells all alike, a group of cells arising by successive generations from the original cell.

This group of cells, pressed one against another, is a spherical, raspberry or mulberry-like mass. It has been proposed to designate this first stage of embryonic development common to all animals by a special name, that of *morula.*

In mammals this solid and compact mass of vitelline cells soon becomes hollow in its center, where a fluid collects, and becomes denser at the surface. The egg is then transformed into a spherical vesicle, whose envelope is constituted by a more or less thick layer of juxtaposed cells, while the interior is filled with fluid. This pouch is called the *blastula,* or the *blastodermic vesicle;* the wall, the *blastoderm,* and its elements, the *blastodermic cells.*

The blastodermic vesicle has a diameter of about 1 millimeter. It is formed by a single layer of cells. At one point, this wall is doubled by a little mass of segmentation cells of elliptical outline, projecting into the blastodermic cavity, simulating on the surface the appearance of a spot called the germinative area, the primitive rudiment of the mammalian body.

The portion of this cellular mass that forms its central limits soon develops actively, making a new layer spreading out on the internal face of the blastoderm, and arranges itself as a second layer. There are thus two layers or two *lamina,* containing between them in the region of the germinative area an intermediate mass. These two leaflets have different characteristics; they are called, the external layer, or *ectoderm,* and the internal layer, or *entoderm,* or also *epiblast* and *hypoblast.* As to the part included between the two sheets in the region of the germinative area, it is the *intermediate mass* or *mesoblast.*

In birds, reptiles, plagiostomes and cephalopods, the insects, the higher arachnidae, and the crustacea, which have eggs with a voluminous nutritive vitellus, there is a *partial segmentation,* involving only the primitive vitellus. Thus these eggs are said to be *meroblastic* or partially fractionated, in opposition to the *holoblastic* eggs of the mammals with total fractionation. But this is a difference without importance, for in the one as in the other the primary result of the embryogenic work is the formation of the *two primary layers.*

Total fractionation is also found among the lower animals with the formation of a raspberry like mass or *morula* and the creation of a two layered pouch provided with an opening. This form constitutes the *gastrula,* with its entoderm and its ectoderm. This is what is seen in sponges, polyps, and worms.

As can be seen, there is a certain similarity in the first stage of embryological development in all animals.

Later on four layers are found; this multiplication results, as Remak showed, from the splitting of the mesoblast into a *musculocutaneous layer* and a *fibro-intestinal layer.* As to the epiblast or ectoderm, it takes the name of corneous layer, or sensory, or sensorial cuticle; the hypoblast, or internal layer, is referred to as intestinoglandular. This division into four layers that characterizes the second stage of embryogenic development is encountered in all vertebrates and most invertebrates, except, among the latter, the zoophytes and the spongiosa, where the process is reduced to the division into the two primary layers.

The cells that constitute each of these layers and their descendants have a particular role in the constitution of the organism. The *corneous* or sensitivocutaneous layer, also called the epiblast, forms the epidermis with its appendages, (hairs, nails, sweat glands, and sebaceous glands) and the central nervous system and spinal marrow.

The *musculocutaneous layer* of the mesoblast, or mesoderm, forms the dermis, the muscles, and the internal skeleton with its bones, cartilages, and ligaments, that is to say the muscular and connective systems.

The *fibro-intestinal layer* of the mesoblast forms the heart, the great vessels, the lymphatic vessels, the blood itself and the lymph, that is to say the vascular system, plus the mesentery and the muscular and fibrous parts of the intestine.

The *internal lamina,* the hypoblast or hypodermis, or the intestinoglandular lamina, provides the epithelial lining of the intestines, the intestinal glands, the lungs, and the liver *(see* Fig. 40).

How are these elements arranged, and according to what design and what plan?

It can be replied that this design and this plan are laid out from the beginning and that if these elements form materials of the same nature and the same location, they receive a distinct architectural destination from the beginning; they serve to erect a monument in a particular style which is revealed and can be predicted as soon as it begins to be executed.

In the vertebrates, from this moment, the germinal disk exhibits two parts; an opaque marginal zone, the *area opaca,* surrounding a clear central region, the *area pellucida.* The most central cells of the external and middle layers multiply in the *area pellucida* and form an oval spot even more brilliant than the germ itself, the *protosoma.* A groove, the primitive furrow, soon divides this germ in two halves and the edges of the groove thicken so as to form protuberant ridges by the proliferation of the cells of the external lamina. The contour of the germ changes at the same time, and constricting near its middle, takes the form of a violin (*see* Fig. 38) . During this time the middle lamina, the mesoderm, thickens and behaves differently in its central part, its peripheral part, and in its intermediate region; its central part, subjacent to the groove, differentiates and begins to organize itself to form the cellular cylinder called the *dorsal cord*; the peripheral part of this mesoblast becomes fissured to constitute the two layers, musculocutaneous and fibro-intestinal, which tends to separate one from the other, leaving a cleft between the two, the rudiment of the coelom or pleuroperitoneal cavity. As to the intermediate zone of the middle lamina, between the dorsal cord in the center and the divided part at the periphery, it constitutes on each side a sort of cord called the *primitive vertebral cord,* from which come the segments of the vertebrae.

The dorsal ridges formed by the external lamina approach each other, meet, and fuse, and thereby form a *medullary tube,* destined to become the spinal cord; this will be pushed toward the interior and enclosed in the spinal canal which surrounds it, and this is formed from the right and left vertebral portions of the middle lamina which will come to meet each other at the midline above and below, and form a tube for it.

Things happen in much the same way with the internal lamina or hypoblast, but more slowly. Reduced for a long time to a single layer of cells, this lamina soon shows along the axis of the germ a depression or groove whose edges meet and constitute finally a complete tube, the intestinal tube.

This is not the place to follow the development of these various parts step by step. It is sufficient for us to grasp the general pattern.

In the vertebrates the type is marked and identified from the beginning, in the sense that there is a primitive groove below which the middle lamina, which has remained undivided, forms an *axial cord,* and everything is symmetrical from one side to another. This division of the germ into two halves by a primitive line indicates the direction development will follow and the part of the family tree to which the animal will belong.

The distinctive characteristics of the various vertebrates, and in a general way of the various groups, appear but gradually, and all the later, the more the adult beings resemble each other. Haeckel has enunciated this law in the following terms:

> The more two adult animals resemble each other in their general structure, the longer their embryonic forms remain identical and the longer their embryos can be mistaken for each other or are to be distinguished only by secondary characteristics.

If we were to summarize the preceding facts and include them in a general formula, we would say with Baër:

> The living being comes from an originally identical cell, the primordial egg, it grows by progressive formation or epigenesis, by the proliferation of this primitive cell which forms new cells, which are differentiated more and more, and collect into cords, tubes, and lamina to form the different organs. This structure proceeds to complicate itself successively, so that the forms become more and more specific as development advances. The most general form, that of the family, is manifest first, then that of the class, then that of order, and so on down to the species.

Development thus follows pathways that are common at first and diverge when it is to lead to different forms. The only question under discussion is to know at what point this divergence begins, because at first there is no differentiation and the primary stages seem to be identical. Most embryologists have thought that what there is in common in a group of animals is always developed sooner in the embryo than what is special, and in consequence when four types of structure are conceived, as Cuvier, Baër, and Agassiz did, then it is natural that four types of development or evolution would be recognized. Baër in particular invoked four embryological processes which were characterized from a very remote stage of development, and brought to their perfected form the germs of the animals of the four families of Cuvier. This system was somewhat premature, and modern embryological observations have contradicted it in many of its parts. Of the four basic types accepted by Baër, there is one, *evolution contorta,* which was finally discarded, while another, *evolutio radiata* could not be accepted without definite reservations. Nevertheless, and in the absence of any other classification of embryological processes, we shall recall here the system of Baër; however imperfect it might be, it affords at least a historical interest, and a framework for the new systems to which the meticulous observations of the modern zoologists will lead us.

Baër thus accepted four types of development just as Cuvier accepted four types of organization. He characterized them by the following names:

1. *Evolutio bigemina;* vertebrates.
2. *Evolutio gemina;* arthropods.
3. *Evolutio contorta;* molluscs.
4. *Evolutio radiata;* radiata.

1. The first type, exhibited by the vertebrates, is the *type with bilateral symmetry*. To characterize its development, Baër used the designation, *evolutio bigemina.* Later, Kölliker in his *Entwickelungsgeschichte der Cephalopoden* (Zurich, 1844) accepted

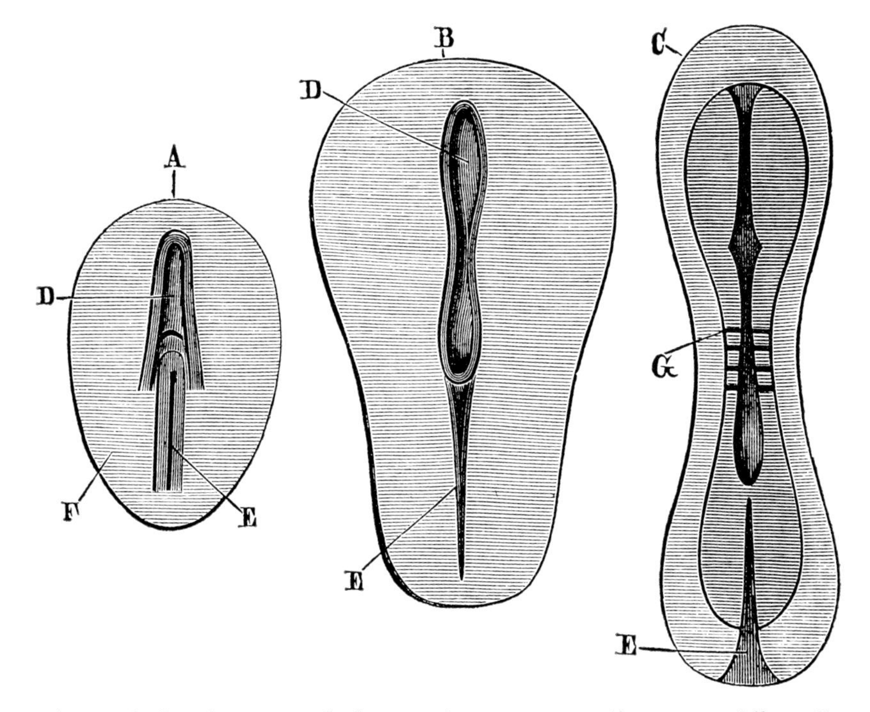

Figure 39. Development of the vertebrates; mammalian type (bilaterally symmetrical development). A, B, C, three stages of the rabbit embryo; D, nervous system; E, axial fillet; F, *area germinativa;* G, primitive vertebrae; Here are seen two symmetrical axes formed one by the nervous system and the other by the visceral system (Heusen and Kölliker).

the same type and the same designation as really expressing the process of development of these vertebrates.

The embryo, arising from a localized portion of the divided egg *(evolutio in una parte)* develops in two different directions, showing bilateral symmetry.

The development of the embryo takes place by a double repetition of the parts; lateral repetition and repetition above and below; that is to say that identical organs are produced which divide on the two sides of an axis (dorsal cord) and project above and below (dorsal and ventral lamina) and meet along two par-

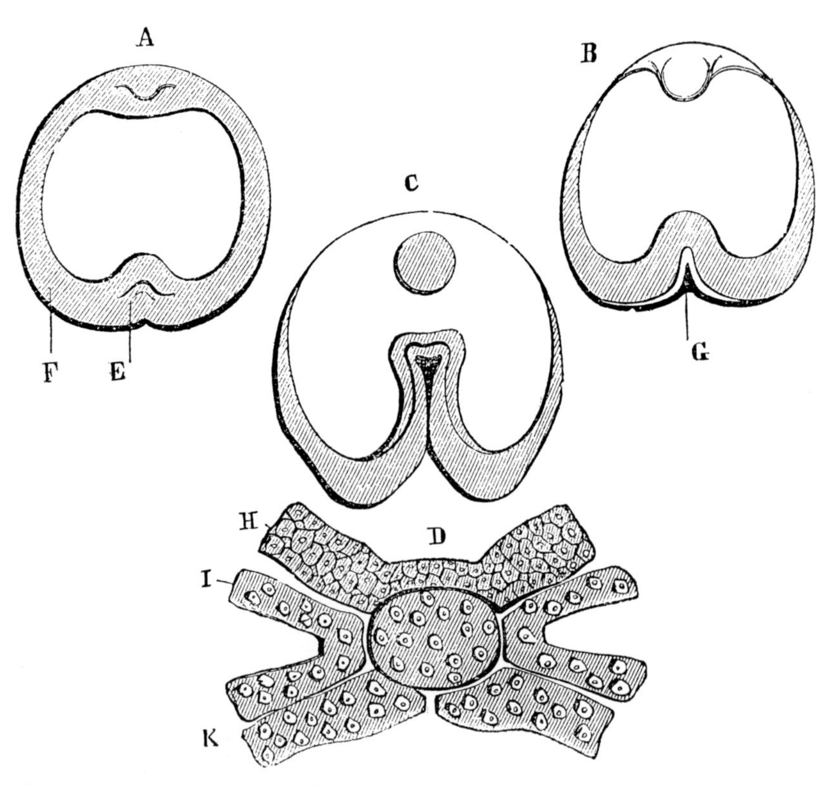

Figure 40. Development of the vertebrates, bilaterally symmetrical development *(evolutio bigemina* of Baër)—Type in fish; A, B, C, three stages of the embryo of the torpedo *(Torpedo oculata);* E, embryo; F, *area germinativa;* G, nervous system; D, section of the embryonic layers; H, ectoderm forming the primitive medulla; I, mesoderm; K, entoderm; in the center can be seen the dorsal cord separating the two axes of development (Al. Schutz).

allel lines so that the internal lamina of the germ closes below and the external lamina above, by which two elongated cavities are formed, the one, the *visceral cavity* that houses and circumscribes the visceral system or vegetative system, and the other, the *medullary cavity* surrounding and circumscribing the spinal cord and brain, the central organ of animal life.

2. The second type of organization and development is exhibited by the articulates (see Fig. 41).

It constitutes the *evolutio gemina* of Baër and of Kölliker. It is characterized by the fact that the dorsal lamina remain open and are transformed into limbs.

Here, development produces indentical parts emanating from the two sides of an axis and fusing along a line parallel and opposed to the axis. This type can also be called the longitudinal type. There is a single cavity which houses the viscera and the nervous system. The intestinal canal, the vascular trunks, and the nervous system extend the length of the body, which exhibits two extremities. It is between these two extremities, a front and back, that the opposition is manifested; it is less clearly exhibited between the top and bottom, because the nervous system extends from one side to another of the digestive system.

The appendicular or dependent parts project laterally, on the right and on the left, as shown on the figures which we place before the reader (see Fig. 41).

3. The third type of organization and development is the least well established of the three and the most likely to undergo radical revision. It is the *massive type,* characterized by the name *evolutio contorta.* It suggests that development produces identical parts, *curved* around a conical or otherwise shaped space. The digestive tube is more or less curvilinear. More complete study of the development of molluscs has established that the coiling shown by some of these animals is not a basic feature anymore than it is general. Moreover, Kölliker himself, at a time that is already distant (1844) considered the molluscs as animals whose development took place uniformly and indifferently in all directions, that is to say that he included them within the type of *evolutio radiata.*

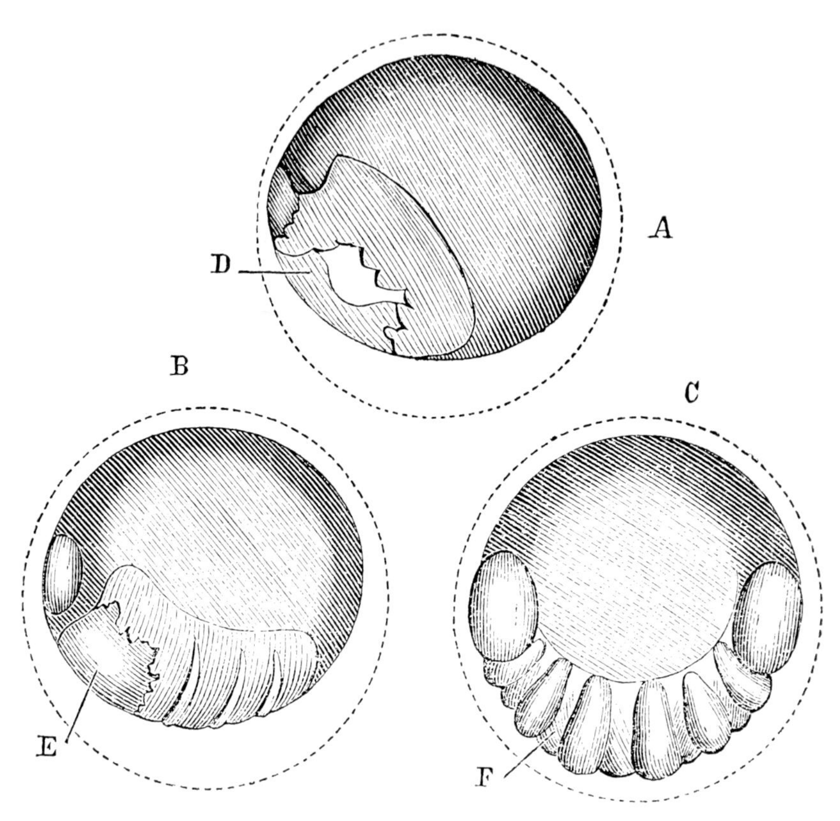

Figure 41. Development of the articulates; example of simple symmetrical development *(evolutio* gemina of Baër) —Egg of an arachnid *(Agelena labyrinthica)* at several stages of development. A, B, in profile; C, full face, D, E, F, symmetrical embryo with respect to a single axis of development (Balbiani).

4. The fourth type of organization and development is represented by the numerous radiates. It constitutes the *peripheric type,* and develops in the way Baër and Kölliker called *evolutio radiata.* All parts of the body arise at once *(evolutio in omnibus partibus).* Development takes place around a center and produces identical parts in a radial order, on a transverse plane. Thus the work of development takes place between the center and the periphery, and it is between these two regions that the essential contrast exists. On the contrary, the contrast is less marked between

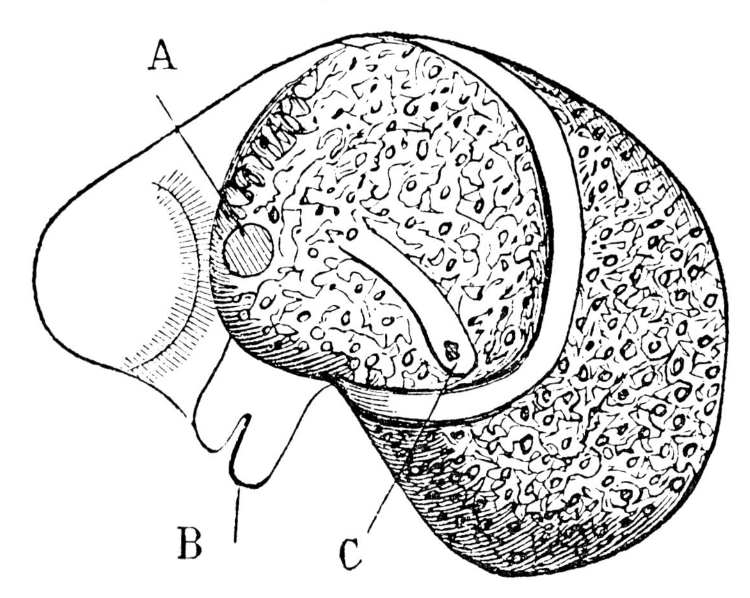

Figure 42. Development of the mollusks; spiral development *(evolutio contorta* of Baër). A young gastropod embryo *(Nassa mutabilis)* seen in profile; A, primitive kidney; B, foot; C, anus, the ending of the terminal portion of the digestive tube which begins behind the foot, describing at the beginning thereby a sharp curve (Bobretzky).

the top and the bottom, parallel to the longitudinal axis, as well as between the front and the back. In consequence the developmental type is one of radiation.

III. ORIGIN AND CAUSES OF MORPHOLOGY

It is particularly through the study of development that one can get an idea of the existence of laws that regulate the morphological constitution of the beings. From the first moment one gets a glimpse of an ideal plan that is given reality step by step; at first one sees the rough sketch, which is perfected and completed in the course of events. The starting point is apparently identical, the end point is infinitely diversified, and the animal goes from the one to the other in a regular and invariable manner by a work that in its complexity is always the same.

If one has only the starting point, when one sees only the

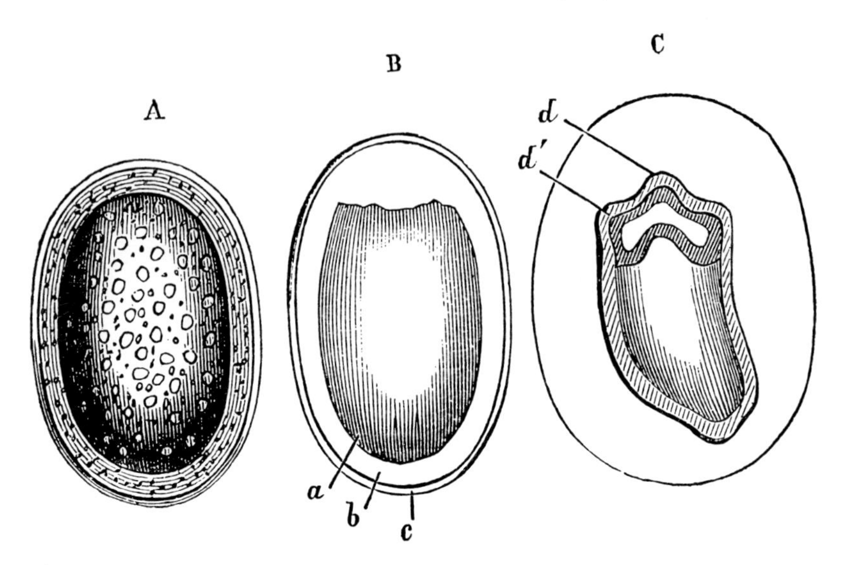

Figure 43. Development of the zoophytes; radial development *(evolutio radiata* of Baër) —A, B, C, three stages of the embryo of a hydra *(Hydra aurantiaca). a,* entoderm; *b,* ectoderm; *c,* membrane of the egg; *d, d',* tentacles presenting their radial appearance from the onset (N. Kleinenberg).

primordial ovule, one knows nothing about what is to happen; one can predict only whether the result of the formative process will be the creation of a zoophyte or a vertebrate, a mammal, or a man.

To predict the outcome of the process the origin of this *protovum* must be known. If one knows from where it comes, one knows what it will become. Thus all morphological development is contained in the previous state. This work is pure *repetition;* it does not have its reasons at each instant in a force currently active; it has its reasons in an anterior force. There is no morphology without predecessors.

In the reality we do not witness the birth of a new being; we see only a periodic continuation. The reason for this apparent creation is therefore not in the present; it is in the past, at the beginning. We cannot find it among the secondary or actual causes; it must be sought in the primary cause.

The living being is like the planet that describes its elliptical orbit by virtue of an initial impetus; all the phenomena that transpire on the surface of this planet, like the vital phenomena within the organism, manifest the play of physical forces that are present and active at this moment, but the cause that impressed on them its initial force is outside these present phenomena and associated solely with the general cosmic balance. It would be necessary to change the planetary system entirely in order to modify it; the present state of affairs is the result of a balance in which all the parts participate, and which would disturb all the parts if it were changed at a single point.

This comparison applies to the living being and its development. Morphology is no more linked to the current vital manifestation than the phenomena of physical agents at the surface of the earth are associated to the movement of our planet on the ecliptic orbit. This is why we separate, absolutely, the phenomena of life, the object of physiology, from organic morphology whose laws are studied by the naturalist (zoologist and botanist) but which escapes us experimentally and are not within our purview.

The law of morphology does not find its *raison d'être* from moment to moment; it exhibits a hereditary or precedent influence which we cannot efface, and a primitive impetus linked to the general cosmic whole, which we are powerless to get at. It results from this that in the present state of things morphology is fixed, and this, it is understood, without regard to any ideas we might form of the evolution that brought it to that point. Whether one is Cuvierist or Darwinist is of little importance; these are two different ways of understanding the history of the past, and the establishment of the present regime; this can provide no means of regulating the future. There can be no changing the ovum of the rabbit to make it forget its primitive impetus and its early stages, and there will not arise from it a dog or some other mammal. The limits between which morphology is fixed, if they are not absolute (there is nothing absolute in the living animal), are at least very circumscribed. If one tries to make a being deviate from its path, as it happens by the creation of artificial varieties, one must constantly maintain it in the new path. Varieties tend ceaselessly to return to their starting point.

It is unnecessary to see in this tendency to return to the starting point any particular mysterious force that watches over the conservation of the species. If things happen in this way it is because the being is in some way imprisoned by a series of conditions which it cannot escape, since they are always repeated in the same way outside and inside it. Thus a carnivor born with the organs of a carnivor, must indeed follow in the direction that these organs give it. It would have been necessary to act before the formation of these organs, and before adult life, but this is impossible because the egg already has the potential of the adult state, and its formation takes place under such rigid conditions that they cannot be changed without causing the death of the beings one attempted to modify. It is thus not astonishing that in similar circumstances the species and types perpetuate and maintain themselves, and one cannot extend experimental intervention beyond certain limits.

In some other cosmic balance the vital morphology would be different. I think, in a word, that there exists in nature an infinite number of potential living forms that we do not know. These living forms would be in some way dormant or expectant, they would appear once the conditions for their existence would develop and once brought into existence they would perpetuate themselves as long as the conditions for their existence and succession would themselves be continued.

It is the same with new substances that the chemists make; they do not create them, they were potentially possible within the laws of nature. The chemist only brings into being, artificially, the external or cosmic conditions for their existence.

The phenomena of development take place, one might say, in consequence of a given initial *cause,* their appearance represents a series of instructions laid out in advance which *in reality* are executed separately. If you see two organs develop successively or simultaneously to work together, so it would seem, for a common goal, you might believe that the influence or the presence of the one logically determined the formation of the other; this would be wrong; the two organs are developed blindly as the result of instructions that might at times seem to be completely illogical, as all instructions are, moreover, when considered in their application to special and unforeseen circumstances. Let us take an example: if the first development of the chick is observed, the heart is seen to form in the cicatricule, and a system of vessels, the *area vasculosa,* spreads out around it and connects with the central circulatory system of the embryo. It would seem quite natural to think that the peripheral vascular system is formed because the heart of the embryo orders it to; not at all. If you prevent the embryo from appearing, the *area vasculosa* is produced none the less, although its function has become totally useless.

We shall make a general remark on this subject which will be de-developed later in more detailed studies. The organs of the body, which are all associated and harmonized in their activity, have their autonomous and independent development. In this respect the organism reflects what takes place in a gun factory, for example, where each worker makes a piece independently of someone else who makes another piece without knowing the whole to which they must contribute. Next there seems to be an assembler who puts all these pieces together harmoniously. In the animal organism it is the nervous system that is the great harmonizer of function in the adult. When this adjustment of the organs in the animal or plant embryo goes away for any reason, it results in the death of the organism or in monstrosities, or *malformations* as they are commonly called.

We wish this essential point to be clearly understood, that the morphology must be completely distinguished from the physiological activity of the organs. The laws of morphology are laws that we have termed dormant or expectant, which neither prevent nor produce any vital phenomenon, which do not act and which cannot be acted upon.

The actual role of the organs is not the agent that has caused their formation. Paul Janet, in his philosophical treatise on final causes.* has assembled all the arguments showing that matters are arranged and harmonized in view of a fixed purpose. We are in accord with him, because without this harmony, life would be impossible, but for the physiologist this is no reason to look for the explanation of morphology in final causes that are presently active. Here as always the category of final causes is confused with the category of initial or primary causes.—Let us take another

*P. Janet, *Les causes finales,* 1876 *(Bibliothèque de Philosophie Contemporaine).*

example. Let us imagine that the development of a certain animal, a rabbit, is followed. The different organs will be seen to be constituted in succession. The eye and its very particular structure is organized precisely to permit the rabbit to receive the impression of light, and according to an adherent of final causes, it is this purpose that will determine its formation and will preside over its fabrication step by step.

It is against this abuse that it is necessary to protest in physiology. The final cause does not intervene at all as a present and effective law of nature. This rabbit might not come to term, its eye will be useless to it; it will never receive the action of light. It is the same in the case of a hen without a male that lays a necessarily infertile egg. The organ is not made with a view to function, for the final cause could be unusually misleading. This would indeed be a blind foresight, whose calculations would so often be frustrated. The eye is formed in the rabbit because it was formed in its antecedents, and because nature repeats its orders into eternity. It is not at all for the function that it will have that nature works. It remakes what it has made; this is the law. It is thus only at the start that its prevoyance can be invoked; it is at the beginning. It is necessary to go back to the first cause. The final cause is the consequence of the first cause; to my view they unite with one another at an inaccessible distance.

The reason why the hen sits on her egg is not actually to produce the development of the young animal. Give her a plaster egg and she will sit on it just the same and complain if it is removed. She sits by virtue of instruction which her antecedents followed, and not for a present purpose or present motive.

Thus we do not accept that the particular forces that are constantly at work in a living being have as their law the welfare of each living being and that it is for this present utility that the divided bile duct reforms and the sectioned nerve repairs and heals itself. To our mind it is wrong to accept in man or in animals as well, an organic force acting to its own best interests with full knowledge of its acts. Aristotle placed a spiritual power (ψυχὴ θρεπτιχα) in each organ, operating outside the *Ego,* unknown to the consciousness, and nevertheless acting in the various circumstances with a perfect discernment. Alexander von Humboldt did not care to decide whether each organic act did not presuppose a force that had conceived it from the beginning in some representative manner.

For us the preexisting law is found only at the beginning, and all there is at the present is its unfolding.

In relating in this way final cause to first cause, the physiologist steps out of his own domain, that is to say from active science to speculative science, to philosophy. Finality is by no means a *physiological law,* it is by no means a *law of nature,* as certain philosophers speak of it; indeed, it is rather a rational law of the mind. Physiologists should guard against confusing the purpose with the cause, the *purpose* conceived in the mind, with the *efficient cause* that is in the object. "Final causes," in the words of Spinoza, "do not at all indicate the nature of things, but only the make-up of the imaginative faculty."

The philosophers who attempt to uproot the principle of final causes from the world of metaphysics and implant it in the objective world of nature adopt a quite different point of view for themselves than men of science. The philosophers start from the axion that what is *real is rational* and that everything that happens is *intelligible.* Things happen, they say, as if the cause of the phenomena had *foreseen* the effect that they would produce. This cause is formed in the image of that which we carry within us, of the volition that presides over our actions. "Having thus within himself a prototype of final cause, man has been led to conceive of it outside *himself,* and as he makes things by art or industry he has imagined that natural objects were made in the same way by art or industry"; this is what the words of Goethe express: nature is an artist. It has been believed that a *thought,* comparable to that of man, directed toward some purpose all the wheels that turn in the organized being, and subordinates to some predetermined future effect, the phenomena that follow each other one by one. Thus, this final effect, in view of which all the phenomena are coordinated, become *retroactively* the directing cause of those that precede it. The *future act* which will appear as a *result,* would become a *purpose* constantly present in the form of ideal anticipation in the series of phenomena that precede it and bring it into being; it becomes a final cause.

This is an essentially metaphysical concept that can be accepted as such.

But the man of science envisages only efficient causes, or efficient conditions, and not according to the expression of Caro,* their *intellectual conditions.* He sees the order, the interrelations of phenomena, their harmony and their *consensus,* he recognizes their predetermined linkages. This is an undeniable fact. The role of science is limited to the establishment of this fact. Janet himself recognizes the right of science to interdict all other studies than those relating effects to their proximal conditions or causes. Without doubt these physical causes or conditions do not suffice to give us a full account of phenomena, but they are enough to make us master of them.

If one wants to determine the primary cause of this vital preordinance one goes outside of science; whether there is an *intelligent and prevoyant intention* as the finalists would have it, or a *condition of existence* as the positivists would have it, a *blind will* according to Schopenhauer, or an *unconscious instinct* as Hartmann says, is matter of opinion. The final cause is one of those interpretations that is adequate for the nature of the mind, invented to gain an understanding of primary causes; it is, according to Caro a *law of reason* or better, the very law essential to human reason confused with the law of *causality.*

But, in thus limiting finality to the domain of metaphysics to satisfy the requirements of reason, it is still necesssary to avoid abusing it. In this category of ideas one can admit, as a physiologist-philosopher, a sort of *particular finality,* or *intraorganic teleology;* the grouping of vital phenomena into functions is the expression of this idea. But in this case the final cause, the purpose, is sought within the object itself but not outside of it. Every act of a living organism has its purpose within the confines of this organism. This forms in fact a microcosm, a little world where things are made one for another, and the relationship can be comprehended because one can embrace the *natural* whole of these things.

*Caro, *Journal des Savants,* 1877.

This particular finality is alone absolute. Only within the confines of the living individual are there laws that are absolutely predetermined. Only there can one see an intention that is carried out. For example the digestive tube of the herbivor is made to digest the food stuffs that are encountered in plants. But the plants are not made for it. There is only one necessity for its life; a necessity that must be obeyed, that is that it nourish itself; the rest is contingent. The relations of animals to plants are purely accidental and not necessary. Nature, it might be said, has made things for themselves and without bothering with contingencies. It does not condemn certain beings to be devoured by others, on the contrary it gives them the instinct of conservation, of proliferation, and means of resistance to escape death. In résumé, the laws of particular finality are rigid; the laws of general finality are contingent.

The concept of particular finalities can be of help to the mind and to the intellect.

On the contrary all extraorganic finality ought to be rejected. To comprehend the relationship of two natural objects in the external world, it would be necessary to apprehend the exterior world in its entirely, the macrocosm in its whole. This is impossible, and will always be so as the limit of human knowledge. Let us add, moreover, that in fact, all the attempts of this kind have led to nothing but ridiculous conclusions, or have fallen before the attack of the most serious doubts.

To return to the beginning of this discussion, physiology points out the existence of laws of morphogenesis, but does not study them at all. These morphogenic laws derive from causes that are out of reach; physiology keeps within its domain only what is within our reach, that is to say, the phenomenal conditions and the material properties by which the manifestations of life can be grasped.

The study of morphological laws constitutes the domain of zoology and phytology. Aristotle considered that in the living being what is the most essential is precisely that form that is so profoundly impressed upon it by a sort of ancestral heritage. Zoology was for him therefore the study of life itself. Today we separate physiology from zoology because we separate vital *phenomenology* from vital *morphology*.

We can hardly even *contemplate* vital morphology since its essential characteristic, heredity is not an element that we have within our power, and over which we are master, as we are of the physical conditions of the vital manifestations. Vital phenomenology, on the contrary, we can direct.

In truth, heredity can be considered to be an experimental condition, and it could be employed, as is done in zootechnology, by crossings and by selection. Thus fugacious atavisms are substituted for fundamental atavism; but in such experiments conditions are set in motion that remain no less obscure. It is, we repeat, this general morphology of the living being, with the particular and independent morphologies of its various organs which constitute the true field of zoology as a distinct science. Establishing its role in this way, that of physiology and the difference between these two branches of human knowledge are established at the same time.

NINTH LECTURE

Résumé of the Course

SUMMARY: I. Concept of Life—Life is neither a principle nor a resultant; it is the consequence of a conflict between the organism and the external world—Demonstration of this proposition by various detailed expositions.

II. Concept of Living Organisms—Life is independent of a definite organic form—Law of formaton of organisms—The organism is formed in view of the lives of the elements—Autonomy of the lives of the elements and their subordination to the whole—Laws of differentiation and of division of labor—Law of Organic Perfectionment—Morphological unity of the organism—Various demonstrations — Redintegration, cicatrization, etc. — Diverse forms of vital manifestations — Vital phenomena — Functions — Properties.

III. Concept of the Science of Physiology—General and descriptive physiology—Comparative physiology—The problem of physiology: to know the laws of the phenomena of life and act upon the manifestation of these phenomena—Physiology is an active science—Its principle is determinism, like that of all experimental sciences.

I. CONCEPT OF LIFE

We have now arrived at the goal that we wished to attain; we have sketched the whole of the phenomena of life, considered them in their greatest generality. Now let us summarize the essential features of this picture.

Let us first see what concept we ought to form about life. We have established from the beginning that it was illusory to seek to *define* life, that is to say to aspire to penetrate into its essence, just as it is illusory to seek to grasp the essence of any phenomenon, whatever it may be, physical or chemical. The various attempts that have been made in the history of science with the aim of defining life have all ended, we know, in considering it either as a particular principle, or as the resultant of general forces of nature, that is to say, in two concepts, vitalist or materialist. The one and the other are ill-founded; the first, the vitalist doctrine, as we have established, because the supposed vital principle would not be capable of doing anything and consequently of explaining anything by itself, and on the contrary would employ the action of general physical and chemical agents. The materialist doctrine is also just as inaccurate, because the general agents of physical nature capable of causing the appearance of vital phenomena in isolation do not explain their regulation, consensus, and integration.

Adopting the point of view of the special activity of the organisms, one might perhaps say that the vital properties are at once resultant and principle. In fact, the higher vital facilities, irritability, sensibility, and intelligence, might be considered as the results of the physicochemical phenomena of nutrition; but it would also be necessary to admit that these faculties become the forms or the principles for direction and manifestation of all the phenomena of the organism of whatever nature they might be.

In any event, considering the question in an absolute manner, it ought to be said that life is neither a principle nor a resultant. It is not a principle because this principle, in some way dormant or expectant, would be incapable of acting by itself. Life is not a resultant either, because the physicochemical conditions that govern its manifestation can not give it any direction or any definite form

None of these two factors, neither the directing principle of the phenomena nor the ensemble of the material conditions for its manifestation, can alone explain life. Their union is necessary. In consequence, life is to us a conflict. Its manifestations result from a close and harmonious relationship between the *conditions* and the *constitution of the organism.* These are the two factors that are found together and so to speak in collaboration in each act of life. In other terms these two factors are:

1. The definite external *physicochemical conditions* which govern the appearance of phenomena.
2. The *organic conditions* or *preestablished laws* that regulate the sequence, the concert, and the harmony of these phenomena. These organic or morphological conditions derive by atavism from earlier beings, and act as the heritage that they have transmitted to the living world of the present.

We have demonstrated the necessity of a conflict or a collaboration of these two kinds of elements by examining the three forms that life takes (Lecture II). According to the more or less close relation between the organic conditions and the physicochemical conditions, *latent life, oscillating life,* and *constant life* can be distinguished. In latent life the organism is dominated by the external physicochemical conditions to such an extent that all vital manifestations can be arrested by them. In oscillating life, if the living being is not as absolutely subject to these conditions, it nevertheless remains so shackled to them that it submits to all their variations; active and lively when the conditions are favorable, inert and sluggish when they are unfavorable. In constant life the being seems to be free, liberated from external cosmic conditions, and the vital manifestations seem to depend only upon internal conditions. This appearance, as we have seen, is only an illusion, and it is especially in the mechanism of constant or free life that the close relationships of the two kinds of conditions are exhibited in the most characteristic manner.

Since life is for us a conflict between the external world and the organism, we ought to discard all the vague concepts in which it is considered as an essential principle. It remains for us solely to determine the conditions and present the characteristics of the vital conflict in a general way.

The vital conflict engenders two kinds of phenomena, which we have called:

Phenomena of *organic creation.*
Phenomena of *organic destruction.*

This division which we have proposed, ought in our mind to serve as the basis for general physiology.

Everything that happens in the living being is related to one or the other of these types, and life is characterized by the union or enchainment of these two kinds of phenomena.

This division conforms to the true nature of things and is based solely on universal properties of living matter, with the exception of the morphological complications of the beings, that is to say of the special shapes into which this matter has been molded.

Eighty years ago Lavoisier had the intuition of these two aspects that vital activity can assume, and of the simple and fruitful classification of the phenomena of life which results from it. He foresaw that as its practical purpose, physiology ought to try to ascertain the conditions and the circumstances of these two kinds of acts, organization and disorganization.

1) The phenomena of *disorganization* or of *organic destruction* correspond to the functional phenomena of the living being.

When a part functions, such as muscles, glands, nerves, brain, the substance of these organs is consumed; the organ destroys itself. This destruction is a physicochemical phenomenon, most always the result of a combustion, a fermentation, or a putrefaction. Basically it is a true death of the organism. It corresponds to the functional manifestations that are revealed to the eye, manifestations by which we recognize life and by which, in consequence of an illusion, we are led to characterize it.

2) The phenomena of *organic creation* or of *organization* are the plastic processes that take place in resting organs and regenerate them. The assimilative synthesis assembles materials and reserves that are to be expended in functioning. It is an internal, silent, and hidden work, without an obvious phenomenal expression.

It might be said that of these two kinds of phenomena those of organic creation are the most particular, and the most special to the living being; they have no analogs outside the organism. Thus the phenomena that we include under the heading of *organic creation* are precisely those that most completely characterize life.

We should also recall that these two kinds of phenomena are divisible and separable only in the mind; in nature they are closely united, they take place in all living beings, in an enchainment that cannot be broken. The two operations of destruction and renewal, the one the opposite of the other, are absolutely connected and inseparable in this sense that destruction is the necessary condition for renewal; the acts of destruction are the precursors and the instigators of those by which the parts are reestablished and reborn, that is to say those of organic renovation. The one of these two types of phenomena which is, so to say, the more vital, the phenomenon of organic creation, is thus in some way subordinated to the other, to the physicochemical phenomenon of destruction. We have the proof of this in studying latent life (Lecture II) ; we have seen that in beings sunk into this state of absolute inertia reawakening or revival begins by the prior reestablishment of the acts of vital destruction. In reviving, the animal or plant so to say begins by destroying its organism, by expending materials that had been placed in reserve beforehand. Creative life exhibits itself only secondarily; it manifests itself only in the presence of death, or of the products of destruction.

This is precisely because the plastic or synthetic phenomenon is subordinated to the functional or destructive phenomenon that we have an indirect means of influencing it experimentally by acting on the latter. The subordination exists, it is well understood, only in its execution, for considered in their relative importance, those that regulate the others and evoke them are the least essential, the least vital.

The distinction that we have established between the phenomena of life provides a natural division of physiology, which ought to propose for itself the study in succession of the phenomena of destruction and then of the phenomena of creation.

In general physiology this division, the only correct one, ought to be substituted, as we have maintained at length (Lecture III) for the division into *animal phenomena* and *plant phenomena,* which have for a long time been opposed one against the other. The separation of natural beings into two kingdoms could be founded only on morphological differences in the phenomena but not on their essential nature. All living beings without exception, from the most complicated of the animals to the simplest of plants present to us the two kinds of phenomena of destruction and of organization, with the same general characteristics.

These two kinds of phenomena can be studied separately, and we have outlined the plan and the general features of this study. In Lecture IV we took up the phenomena of organic destruction, which we brought together under three types, namely fermentation, combustion, and putrefaction.

As to organic creation, it has, so to speak, two stages. It includes *chemical synthesis* or formation of the immediate principles of the living substance; in a word the formation of protoplasm; and in the second place *morphological synthesis,* which unites these principles in a particular mould, with a distinct form or shape, which is the shape or the specific design of the different beings, animals and plants.

But this latter synthesis concerns forms that are in some way accessory to the phenomena of life; it is not absolutely necessary to its essential manifestations. Life is in no way tied to a fixed and definite form, it can exist when reduced to the destruction and the chemical synthesis of a substratum which is the physical basis of life, the protoplasm. The concept of morphology is thus, as we established in Lecture V, a complication of the concept of life. At its simplest level (whether realized in isolation in nature or not), stripped of the accessories that mask it in most beings, life, contrary to the idea of Aristotle, is independent of any specific form. It resides in a substance defined by its composition and not by its structure, the *protoplasm.*

After having pointed out the opinions that have been held about this substance, we took up the problem of its creation or formative synthesis.

It is this life, without characteristic forms, properly speaking, whose mechanisms, properties and conditions are common to all beings, which constitutes the real domain of general physiology. The machinery of all living organisms represents only the various aspects of a single unique substance, the depository of life, identical in animals and plants, the protoplasm. It is here that the two types of vital manifestations are located, destruction on the one hand and organization or creative synthesis on the other. In Lecture VI we traced the outlines of our knowledge relating to the synthetic role of the protoplasm, and with this we completed the rapid review of life considered in its universal features, that is to say, traced out the design of general physiology.

In résumé, protoplasm is the organic basis of life. It is between it and the external world that takes place the vital conflict that characterizes life for us, and which we must study and master. But protoplasm, however elementary it may be, is no longer a purely chemical substance, a simple immediate chemical principle, it has an origin that eludes us, it is the continuation of the protoplasm of an ancestor.

We cannot act upon the manifestations of this general life, an attribute of the protoplasm, except by regulating the physicochemical agents which enter into conflict with the preexisting protoplasm. The exact determination of these material conditions is what we have called *physiological determinism,* which is in reality the only absolute principle of the science of experimental physiology.

Such is the concept that permits us to understand and analyze the phenomena of living beings, and affords us the opportunity to act upon them.

II. CONCEPT OF LIVING ORGANISMS

In the living being we have distinguished *matter* and *form.* Study of complex beings shows us that the vital conflict is basically identical at all times, and that comparative physiology is definitely the study of the superficial forms, so to speak, of life, while general physiology comprehends the study of its fundamental conditions.

The living matter, independent of all form, amorphous or rather *monomorphous,* is the *protoplasm.* In it reside the essential properties of *irritability,* the starting point and the rudimentary form of sensibility, the faculty of *chemical synthesis* which takes in ambient substances and creates organic products, in a word, all those attributes whose vital manifestations in the higher creatures are nothing but the diverse expressions of special modalities.

Nevertheless, protoplasm is not yet a *living being;* it lacks a form that characterizes the distinct being, it is the ideal *matter* of the living being, or the *agent of life,* it exhibits to us life in the state of nakedness, in whatever is universal and persistent throughout its varieties and forms.

The *form* that characterizes the being is not a consequence of the nature of the protoplasm. In no way can a property of the latter explain the morphology of the animal or the plant. Form and matter are independent, and distinct, and as we have said (Lecture VIII), chemical synthesis, which creates the protoplasm, must be separated from the morphological synthesis that shapes and models it.

But this independence is dominated by the exigencies of the vital conflict, which must always be respected. From this point of view there must be a necessary relation between the *substance* and the *form* of living beings, and this relation is expressed by what we call the *structural law of organisms*. The structure of these complex edifices that constitutes the animal or plant species depends in a general way on the conditions for the existence of the living matter or the protoplasm. These conditions of protoplasmic activity are taken into consideration in the law of morphology which respects them and uses them, so that in a certain fashion, morphology is subordinate to the elementary vital conditions of the protoplasm, that is to say, to elementary life. This subordination is expressed precisely in the law of structure of organisms which is stated in this way: *The organism is constructed with the life of the elements in view. Its functions correspond fundamentally to the provision in kind and in degree of the four conditions of this life: humidity, heat, oxygen, and reserves.*

The simplest of the forms in which living matter can exhibit itself is the *cell*.

The cell is already an organism: this organism can, by itself, be a *distinct being* (see Lecture VIII), or it can be the individual element of which the animal or the plant constitute a society.

Whether it is an *independent being* or an *anatomical element* of higher beings, the cell is thus the simplest living *form;* it presents to us the first stage of morphological complication, and it can be said that it is at this stage that the protoplasm is put to work to constitute complex beings.

We have spoken at length about the origin of the cellular formation, treating its general morphology in the preceding lecture. It is found to be endowed to the highest degree with all the vital properties encountered already in the protoplasm, namely, movement, sensibility, nutrition, and reproduction.

Form constitutes a new character for it. Form gives evidence of a hereditary or atavistic influence, whose existence, already appreciable in protoplasm, will become altogether outstanding in the higher organisms. We have said that the protoplasm itself is an *atavistic substance,* that we do not see it born, but that we only see it continue (Lecture VI). This hereditary influence is exhibited still more in the cell but nevertheless it is not as much as we shall find it when we examine more and more complicated animals. In fact, form is less fixed in the heredity of a cell than is the form of a complex being in the heredity of this being; there is a certain *cellular polymorphism,* a certain *variability of cellular species,* and the history of histogenesis and of embryological development provides us more than one example of these transformations or of these passages of cellular forms from the one into the other. The observations of Vöchting on the propagation of plants by slipping provide us a striking example of this polymorphism by showing that a cell or a group of cells from the generative zone can, according to circumstances that are wholly within the hands of the experimenter, provide now the tissue of a root, and again that of a bud. The imprint of heredity is the more deeply embedded when it is applied to a more complex being, as though this complexity were the proof of a more ancient origin, or a series of acts more frequently repeated and having, by this very fact, all the more tendency to repeat itself anew.

Now let us look at the higher beings.

The complex organism is an aggregate of cells or elementary organisms in which the conditions for the life of each element are respected and in which the *functioning* of each is nevertheless subordinated to the whole. There is thus at once *autonomy of the anatomical* elements and *subordination of these elements* to the morphological whole, or in other terms, of the lives of the parts to the life of the whole.

We therefore ought to examine in turn the mechanisms by which these two conditions of autonomy of the anatomical elements and of their subordination to the whole are brought about

In a general way we can say that the element is autonomous in that it possesses within itself, and by virtue of its protoplasmic nature, the essential conditions for its life, that it does not borrow nor extract in any way from its neighbors or from the whole; it is moreover associated with the whole by its *function* or by the *product* of this function. An analogy will make our thought better understood. Let us consider the complex living being, animal or plant, as a city having its own special character, which distinguishes it from all others just as the morphology of an animal distinguishes it from all others. The inhabitants of this city represent the anatomical elements within the organisms; all of these inhabitants live in the same way, feed themselves, breathe in the same way, and have the same general features, those of man. But each has his own trade or industry, or aptitudes or talents by which he participates in the life of the society and by which he depends on it. The mason, the baker, the butcher, the industrialist, and the manufacturer provide diverse products, all the more varied, numerous, and differentiated as the society in question has attained a higher level of development. Such is the complex animal. The organism, like the society, is built in such a fashion that the conditions for the life of the elements or the individuals are respected therein, these conditions being the same for all, but at the same time each member depends in a certain measure, by its function and for its function on the *place* that it occupies in the organism, in the social group.

Life is therefore common to all its members, only the function is separate. That which is connected to life itself, that which forms the object of general physiology, is identical from one end to the other of the organic kingdom, and every time a fact of this order has been discovered under the conditions of particular experiments, it is legitimate to extend it.

Until now the general laws of the organization have not been clearly established. Two attempts have nevertheless been made to explain the formation of the complex or higher beings. These attempts are expressed by the law of *differentiation* and by the law of *division of labor*. We shall state shortly why the principle that we propose under the name of the *structural law of organisms* seems to us to be more in keeping with the true nature of things.

We have said that the living organism is an association of cells or elements, more or less modified and grouped into tissues, organs, apparatus and systems. It is thus a vast mechanism resulting from the assemblage of secondary mechanisms. From the monocellular being to man, all degrees of complication are encountered in these groupings; organs are added to organs, and the most highly developed animal possesses a great number of them that form the circulatory system, the respiratory system, the nervous system, etc.

It has been believed for a long time that these superadded mechanisms had their own raison d'être or that they were the result of the caprice of an artistic nature. Today we ought to see in them a growing complexity regulated by law. Anatomy, restricting itself to the observation of forms, did not succeed in deriving it. It is physiology alone that can give an account of it.

Organs and systems do not exist for themselves, they exist for the cells, for the innumerable anatomical elements that form the organic edifice. The vessels, the nerves, the respiratory organs appear as the histological framework becomes complicated, so as to create around each element the environment and the conditions that are necessary for this element, so as to dispense to it in appropriate measure the materials that it needs; water, food, air and, heat. In the living body these organs are like the factories or the industrial establishments in an advanced society which provide the various members of this society with the means of clothing, heating, feeding, and lighting themselves.

Thus the *law of the structure of organisms* and of *organic development* is bound up with the law of *cellular life*. It is to make possible and to regulate more closely the life of the cells that organ is added to organ, and apparatus to systems. The task imposed upon them is to bring together in quality and quantity the conditions for the life of the cells.

This task is absolutely indispensable; to accomplish it they go to work in different ways, they divide the labor, more numerous when the organism is more complex, less numerous when it is more simple, but the purpose is always the same. One might express this condition of organic improvement by saying that it consists in a *more and more extensive division of labor in preparing the composition of the internal environment.*

Thus differentiated and specialized, the anatomical elements live their private lives in the place they are assigned, each according to its nature. The action of poisons, which bear primarily on this or that element while sparing this or that other one, as I have shown for curare or carbon monoxide, is one of numerous proofs of this autonomy. The anatomic elements *behave in association as they would behave in isolation* in the same environment. It is in this that the *principle of the autonomy of the anatomical elements consists;* it affirms the identity of free and associated life on the condition that the environment is identical. It is by the mediation of the interstitial fluids, forming what I have called the *milieu interieur,* that the solidarity of the elementary particles is established, and that each one receives the repercussions of the phenomena that take place in the others. The neighboring elements create for the one under consideration a certain ambient atmosphere, and it feels the changes in it, which regulate its life. If one could create at each instant an internal environment identical with the one which the action of neighboring parts are constantly creating for a given elementary organism, this one *could live alone exactly as in society.*

Subordination of the Elements to the Whole

But this condition of the identity of the environment is indeed restrictive. In the present state of our knowledge it would be impossible to create artificially the *internal environment* in which each cell lives. The conditions of this environment are so delicate that they escape us. They exist only in their natural place, which the accomplishment of the morphological plan assigns to each element. The elementary organisms do not encounter them except in their place, at their station; if they are transported elsewhere, and even more importantly if they are removed from the organism, by this very fact their environment is changed, and in consequence their life is changed or even made impossible.

It is by the infinite variety that the internal environment presents from one place to another and by its particular and constant composition at a given point that the subordination of the parts to the whole is established.

Several examples will give an understanding of the conditions of associated life, where each element is at once free and dependent.

It is now known that bones are formed and renewed thanks to cellular elements in the internal layer of the periosteum. The surgeons have made use of this idea in their practice.

If a sheet of periosteum is selected and removed, and if, taken from its environment it is carried to some other organic region, it will be seen to develop in this unusual place and give rise to a new bone.

In the rabbit and guinea pig, for example, fragments of bone taken from some part of the skeleton have been made to develop under the skin. The property of secreting bony material does not therefore reside in this or that fixed architectural region of the living being; it resides in the cells of the periosteum which take it with them and retain it wherever they might go.

But this autonomy has been exaggerated, and the rights of the total organism, for which the activities of the cells are harmonized, have been ignored. Following the development of this new bone, it has not taken long to notice that it does not survive indefinitely; it is reabsorbed and disappears after a certain time. It has not continued to live under conditions that were not made for it. The periosteal cells that had already been formed have continued the development they began and ended in the formation of bone, but new ones have not been formed at all. The periosteal transplant has disappeared.

This experiment can be given a still more striking form. In a young rabbit an entire bone is taken out of one of the paws, a metatarsals, it is introduced under the skin of the back and the wound closed. The displaced bone continues to live, it even carries out its own development, it grows a bit and the ossification of its cartilagineous parts continues, but soon the development stops, resorption begins to be manifest, and there is no outcome other than the complete disappearance of the transplanted bone.

On the contrary, in the metatarsal space which had been made vacant, a new bone appears and persists, taking the place of the bone that was removed, because the appropriate site is found there.

The experiments on the regeneration of bone which have been invoked to show the absolute autonomy of the anatomical elements have thus ended in a contrary result, in so far as they give proof at the same time of the restrictions placed upon this autonomy. They have revealed the influence that the *place* of the element in the total plan exerts on its functioning. There is thus another condition that does not relate to the element itself, but relates to the morphological plan, to the whole organism. The cell has its own autonomy which permits it to live, as far as it is concerned, always in the same manner, and wherever appropriate conditions are brought together; but on the other hand these appropriate conditions are fully realized only in special places, and the cell functions differently, works differently, and undergoes a different development according to its place in the organization.

Redintegration

Subordination, the condition restricting the autonomy of the elements, is more or less pronounced. The less elevated the organism, the weaker is the chain of subordination among the whole of its parts.

In plants the subordination of the parts to the whole, which in some way expresses the rights of the organism, is at its minimum. A part of a plant can be removed and taken elsewhere to start a new plant to develop. It is on this fact that the practice of grafting and slipping is based. A cell from the bark, for example, can become a bud and replace a cut branch. This change takes place in the cells under the influence of the fluids from the branches whose composition has been modified by the cut.

In animals healing of a wound also takes place under similar influences.

It is the subordination of the parts to the whole that makes an integrated system, a whole, or an individual, out of the complex being. It is in this way that *unity* is established within living bodies. Unity, as we have just said, is the least marked in plants. Among the lower animals also, isolated particles can live when they are separated from the rest of the organism, as occurs in hydra and planaria.

Dugès and de Quatrefages have carried out some interesting experiments on planaria (Figs. 44 and 45). They cut one of these worms in two parts; the one anterior, the other posterior; each of these completed itself and reconstituted a new planaria. One of these animals can be quartered or cut in eighths; there forms just so many new individuals as there are pieces.

It is also known that when working with lizards and salamanders a tail or limb that is cut off can be caused to reappear. An Italian physiologist has carried out some interesting studies in this regard; he has noted that the weight of the animal did not change sensibly during the regeneration. Vulpian has observed similar facts in the tadpole. The same thing happens when a planaria is cut in two; each of the new planaria is and remains quite small. The formation of the new being does not seem therefore to be a true new organic creation recommencing a disrupted activity, but simply the continuation of a development which follows its course by virtue of an original impetus.

We do not need to multiply examples of regeneration at this time; we will recall only those of Philippeaux on the regeneration of limbs in the salamander. A paw removed from the animal will be regenerated: the development of cells in the stump is directed in such a way that the lost member is remade. The new formation that tends to reestablish the integrity of its organic plan manifests quite evidently the influence of the whole on the development of the parts. But it is not even the whole organism that extends its power this far. If the base of the member is removed, regeneration does not take place. The base is like a sort of collet or germ, comparable to the bud which during embryological development had contributed in the very same way to the production of the limb.

From all these examples it emerges that each part develops in such a way as to complete the design of the whole animal. The organism considered as a whole or as a unit thus intervenes and manifests its role through that power of regeneration which per-

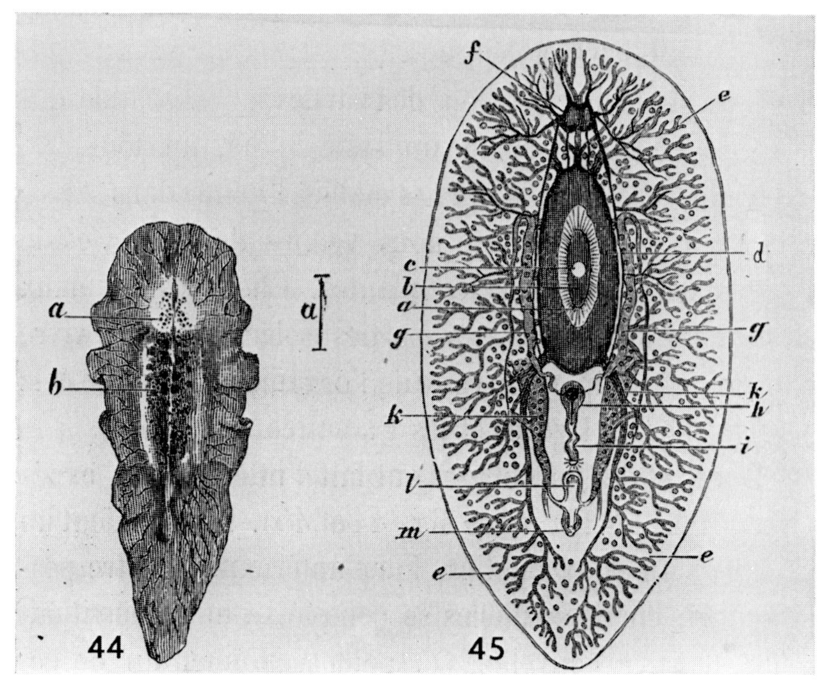

Figure 44.—*Common planaria.* aI natural size.

Figure 45.—*Anatomy of the common planaria. a,* mouth; *b,* proboscis; *c, cardiac* orifice; *d,* stomach; *e, e, e,* gastrovascular ramifications; *f,* brain and nerves; *g, g,* testicles; *h,* seminal vescile fused with the penis; *i,* canal of the penis; *k, k,* ovicular bodies; *l,* copulatory pouch; *m,* orifice of the organs of generation dispersed throughout all the spaces of the body (Edwards, Quatrefages and Blanchard, *Recherches Anatomiques et Physiologiques Faites Pendant un Voyage sur les Côtes de la Sicile.* Paris, 1849).

mits it to repair itself and maintain itself anatomically and physiologically.

An essential remark must be made concerning the accomplishment of the phenomena by which the organism repairs and reestablishes itself. These phenomena seem to have the ability to manifest themselves only when the parts are in their natural place, when they have not been dissociated, as though each of them resulted from a general consensus of all the parts. When, thanks to artificial respiration and circulation we work on organs or parts separated from the organism, we obtain only partial phenomena, of the nature of phenomena of organic decomposition, but the *phenomena of organic synthesis* can no longer be obtained. When physiologists examine an isolated muscle, for example, they can observe all the functional acts, the contraction of the muscle, and the phenomena that derive from it, but the muscle does not nourish itself, does not regenerate itself any longer, and from then on can only use itself up. For this reason the persistence of functional life can only be transitory.

In spite of all the reservations that we have just pointed out, the principle of autonomy of the anatomical elements can be considered as one of the most fruitful in modern physiology. This principle, or under another name, this cell theory, is not a vain expression. It has been a mistake to forget it when concerned with complex organisms. Then one speaks of organs, tissues, apparatus, and puts aside completely the ideas that are associated with the cell.

The cells should never be lost from sight; they are the elementary materials of the whole organism, their life, basically always the same, results from a conflict with the physicochemical conditions of which the experimenter is the master. It is in this way that he can get at the whole being. Every modification of the organism goes back ultimately to an action exerted on a cell. This is a law that was formulated from the first time in my *Leçons sur les Substances Toxiques.** All phenomena, physiological, pathological or toxic, are basically only general or special cellular actions.

The anesthetics, for example, influence all the elements because they act upon the protoplasm, which is common to all. Most poisons influence only special elements, because they act on the products of differentiated cells. Example: carbon monoxide, which acts on hemoglobin, and curare, which doubtless acts on some organic mechanism at the terminals of the nerve in the *muscle.*

In résumé, life resides within each cell, in each organic element, which functions on its own account. It is not centralized in any part, or in any organ or apparatus of the body. All these ap-

*Cl. Bernard, *Leçons sur les Effets des Substances Toxiques et Médicamenteuses.* Paris, 1857.

paratus are themselves created in the light of the life of the cell. When in destroying them, the death of the animal is caused, it is because the lesion or dislocation of the mechanism have eventually had repercussions on the elements, which no longer receive a *milieu exterieur* appropriate to their existence. What dies, just as what lives, in the last analysis, is the cell.

Everything is made by the anatomical element and for the anatomical element. The respiratory apparatus brings oxygen, the digestive apparatus introduces the foods necessary for each one, the circulatory apparatus, and the secretory apparatus make certain of the renewal of the environment and the continuity of the nutritive exchanges. The nervous system itself regulates all these mechanisms and harmonizes them from the standpoint of the life of the cells. The fundamental apparatus essential to the higher organisms all act therefore, the nervous system included, to provide for the cells the physicochemical conditions that are essential to it, the most general of which we have pointed out above.

In this life of cells in association that constitutes the morphological ensembles or living beings, there is at once both autonomy and subordination of the anatomical elements.

The autonomy of the elements and their differentiation explain the variety of the vital manifestations. Their subordination and solidarity make us understand their concert and harmony.

Various Forms of Vital Manifestations: Vital Phenomena, Functions, Properties

The cell is a virtual image of a higher organism. It possesses a general property, irritability. By this abstract or metaphysical expression we mean a concrete, objective *fact,* namely that the phenomenal manifestations that take place within it, such as nutritive exchange, mobility, etc., appear as a reaction provoked by external excitants.

Considering those beings with higher organization, their vital manifestations result in the last analysis from the manifestations of cells, exaggerated, developed, and concerted one with the others. In those complex phenomena that we are going to see in the higher beings, *acts* and *functions,* there are thus two factors;

specialized *cellular activities* and a *concert* among the cellular activities, which directs them toward a predetermined result.

Let us examine these two points.

As the living being advances and is perfected, its cellular elements differentiate more and more; they specialize through the exaggeration of one of their properties to the detriment of the others. Life in the higher animals is more and more distinct in its manifestations; it is more and more diffuse in the lower animals. The vital manifestations are more isolated and more clear-cut at the higher levels of the scale than at its lower levels, and this is why the physiology of the higher animals is the key to the physiology of all the others, contrary to what is generally said.

The properties of the elements are exaggerated in the tissues, as we have just said, by a true specialization. Isolated cells, and monocellular beings can utilize fatty, starchy, and albuminoid foodstuffs which they find in the ambient environment. In the higher animals this property of digesting (by means of ferments, of cellular products) is exaggerated in certain cells brought together to form the pancreatic gland, for example, and these work for the organism as a whole. In résumé, progressive specialization takes place by exaggeration of one property in the cells of tissues and organs.

The phenomena of life include events of growing complexity such as properties, acts, and functions. The property, as we have said, belongs to the cell, at least in the rudimentary state; it is in the protoplasm as an elementary principle, such as contractility. The name *property* is not experimental; it is already abstract, metaphysical. As we have already said, it is impossible to speak except by making abstractions. In the present case the form of language does not mask reality in any profound way, and in the name we can always perceive the fact that it expresses. Under the name of contractility, for example, we perceive the fact that protoplasmic matter changes its shape and form under the influence of an external excitant. And as this feature is not *at present,* at least, reducible to another more simple, and is not explicable by any other, we say that it is distinctive, special, or particular, and we call it a *property*.

Thus, in résumé, the property is the name of a *simple fact,* abstract, as Chevreul said, and at present irreducible; the property belongs to the cell, and to the protoplasm.

Actions and functions, on the contrary, belong only to organs and apparatus, that is to say, to ensembles of anatomic parts.

Function is a series of acts or of phenomena, grouped and harmonized in view of a predetermined result. In the execution of a function the activities of a multitude of anatomical elements intervene, but the function is not the simple sum of the elementary activities of juxtaposed cells; its component activities continue from the one to the other, they are harmonized, and concerted so as to work together towards a common goal. It is this result, foreseen by the mind, which produces the association and the unity of these component phenomena, that constitute the *function.*

This higher consequence, toward which the efforts of the cells appear to work, is more or less apparent. There are thus functions that all naturalists accept and recognize; circulation, respiration, and digestion. There are others about which there is no agreement at all.

There cannot, in fact, fail to be a certain arbitrariness in a decision in which the mind participates to such a great degree; it is the mind that grasps the *functional relationship* of the elementary activities, which proposes a plan, a purpose, for things that it sees happening, that perceives the realization of a result whose necessity it foresaw. But agreement can exist about the well established *material fact* and never *about the idea.* From this arises the disagreement and the differences among physiologists in the classification of functions.

Of completely objective vital functions, completely real, and as independent as possible of the mind that observes them, there are only the *elementary phenomena.* Once one arrives at the concept of a harmony, a grouping, a whole, a purpose assigned to multiple efforts, or a result toward which the actions of the elements tend, then one leaves objective reality, and the mind intervenes arbitrarily from its own points of view–Apart from the intervention of the mind, and insofar as there is objective reality, there is in the organism only a multitude of acts, of material phenomena, simultaneous or successive, dispersed among all the elements. It is the mind that grasps or establishes their interconnections and their relationships, that is to say, their function.

Function is thus something abstract which is not represented materially in any of the properties of the elements. There is a respiratory function, a circulatory function, but there is not anything in the contractile elements that contributes to a circulatory property in them. There is a vocal function in the larynx, but there are no vocal properties in its muscles, etc.

The practical conclusion from these considerations is that it is necessary above all to know objectively the fixed and invariable elementary properties which are the fundamental basis of all the manifestations of life. This is the goal that *general physiology* proposes for itself.

Life lies truly within the organic elements, it is there that we ought always to place the real physiological problem, which is revealed by the action of physiologists on the phenomena of life. It is by determinism applied to the knowledge of these organic elements that we can succeed in approaching the phenomena of life, but never by acting on properties, functions, and on life itself, all metaphysical concepts. We have said often enough that we act directly only upon the physical, and on the metaphysical only in an intermediate way.

We have mentioned above the endeavor to give an account of the conditions for the growing complication of organized beings, from the simple to the most complex forms, by means of two general principles, the principle of differentiation and the principle of the division of physiological labor. We ourselves propose a third point of view that we shall express in our law of the structure of organisms.

Successive *differentiation* is a demonstrated fact, when the development of a given being is followed. Embryological studies, since C. F. Wolff, have established that the animal is formed by *epigenesis* (Lecture VIII), that is to say by the addition and successive differentiation of parts.

When it becomes a question of comparing the various beings, in so far as it concerns the elementary organisms, or the elements, we accept the reality of this law. We have said in fact that we find in a rudimentary state in the cell and in its protoplasm those general properties that are amplified or progressively specialized in the different cells. The cellular elements, we have said above, are differentiated and become specialized by an exaggeration of one of these properties to the detriment of others, and we have given examples of it.

This differentiation, this specialization, is, in sum, a division of physiological labor; an incomplete division, since each element, manifesting one property in an exaggerated way, naturally possesses the others without which it could no longer live.

Within these limits, and with this restriction, the principle of the division of physiological labor seems to us to be correct; it is an expression of the truth.

Apart from this, it is most often applied in an improper and erroneous manner. In a word, the principle is true in general physiology, but subject to error in comparative physiology. It presupposes in fact that all organisms carry on the same activities, with more specialized instruments and more perfectly at the top, and with fewer instruments and with less specialization at the bottom of the animal scale. But this is not true except for the *vital work* that is truly common to all beings; that is to say for the conditions essential to elementary life; it is not true for the functional manifestations that are not necessarily common to all beings. An additional organ does not imply the idea of a more perfect instrumentation in the service of the same requirement; it implies a new work, a new complication of the work. Passing from a white blooded animal with gills to one with a trachea or lungs, one would not understand an application of the law of division of labor, since these organs are separate mechanisms, not performing the same labor at all.

On the contrary, whenever the principle of the division of labor has been denied in general physiology, or indeed when it has been affirmed too rigorously, without taking into account the restrictions mentioned above, it has been a mistake. Thus the dualistic theory that we rejected (Lecture V) is a product of this doctrine. The vital work of the elements, including creation and organic destruction, that belongs to all beings, is by the dualist doctrine divided between two groups of beings, the animals on the one hand and the plants on the other. To the latter belong the organic synthesis of immediate principles, to the other the destruction of these principles. We have seen that this is an error.

The principle of the structure of organisms that we have just presented does not seem to us to be subject to these reservations and these restrictions.

III. CONCEPT OF THE SCIENCE OF PHYSIOLOGY

Physiology, we have said, is the science that studies the phenomena pertaining to the living animal, but taken in this way, the science is still too vast and ought to be subdivided into general physiology and descriptive physiology, either special or comparative.

General physiology gives us an understanding of the general conditions of life that are common to living beings universally. There we study the vital conflict itself, independent of the forms and mechanisms by which it is manifested. Descriptive physiology, on the contrary, gives us an understanding of the form and the special mechanisms that life employs to manifest itself in a given living being. If now one wishes to compare the form of these diverse mechanisms, infinitely varied among the living beings, in order to deduce the laws of these phenomena, this would be the task of comparative physiology. This is of great interest to us because it shows us the infinite variety of life that is based upon the constancy of its conditions; this is what general physiology gives us, and it is here that we must go if we wish to understand the vital motor itself.

If we were permitted a comparison, we would turn our mind back over the numerous applications of steam in industry, and the infinite number of machines it activates. The study of these machines includes a general part and a special part. It is necessary to know the properties of steam, the conditions for generating it, its expansion, the power it develops, and its condensation. This first study corresponds to general physiology, when it concerns living machines. On the other hand, it is necessary to know the particular application that is made of it in the machine in front of us. For this it is necessary to understand the machinery, to know its organs, and so to speak, to master its anatomy. This second study corresponds to special or comparative physiology when considering all the living machines.

Among all these machines there is therefore something identical and something different. It would be dangerous for the engineer to transfer conclusions from one to the other if he kept only the general properties in mind; he can make no legitimate conclusions unless he envisages the particular arrangement of the working parts as they vary from one to another.

It is the same with the physiologist; he can extend his conclusions from animals to man, from one animal to another, and even from animals to plants in all that concerns the general properties of life. He can say nothing about special mechanisms. An example will serve to fix our thoughts. When the facial nerve is cut on both sides in the horse, the animal soon dies of asphyxia. If, carrying the experimental result in the horse to man, one were to say that bilateral facial paralysis would also cause death, one would make a mistake, because after this paralysis the man has lost only the mobility of his facial features, but he continues to breathe and carry out all his vital functions. Nevertheless the general properties of the facial nerve are the same in the horse as in man, but the facial governs different mechanisms in the two cases. One cannot make legitimate conclusions when it is a question of comparing the disturbances that result from the disruption of these mechanisms, but one can conclude, on the contrary, on the identity of the nerve that governs them.

In a word, it is necessary to distinguish between the *properties* that belong to the *elements* and are taught by general physiology, and the *functions* that belong to *mechanisms* and are taught by descriptive and comparative physiology. What pertains to properties can be generalized, but what concerns functions cannot be, except conditionally and after examination.

Physiology should propose for itself the same problem as all the experimental sciences.

Science has *action* as its definitive purpose.

Descartes has already said it: "Knowing the force and actions of fire, water, air, the stars, and the skies, and of all the other bodies that surround us, we ought to be able to employ them for all the uses for which they are appropriate, and thus make ourselves masters and possessors of nature."

The cartesian conception of vital organization made it possible to extend this domination over the vital phenomena because they obeyed physical laws. "I feel sure," said Descartes, "that (by understanding medicine better) one would be exempt from an infinity of ailments, both of the body and of the mind, and perhaps even of the debilities of old age."

The goal of all sciences, as much of the living beings as of inanimate objects, can be characterized by two words; *foresee* and *act*. This is in fact why man has persisted in the painful search for scientific truth. When he finds himself in the presence of nature he obeys the law of his intelligence by trying to anticipate or master the phenomena that erupt around him. *Anticipation* and *action,* this is what characterizes man as he faces nature.

In the physicochemical sciences man goes forward in the conquest of inanimate nature, of dead nature; all the terrestrial sciences whose object can be attained are nothing more than the rational exercise of domination of the world by man.

Is it the same with physiology as with the other sciences? Can the science that studies the phenomena of life hope to master them? Does it propose to subjugate living nature just as nonliving nature has been subjugated? We do not hesitate to reply in the affirmative.*

Physiology ought therefore to be an active and conquering science in the manner of physics and chemistry.

But how can one act on the phenomena of life?

Having come to the end of our study, here we are again facing the problem of physiology, as we posed it at the beginning. The phenomena of life are represented by two factors; *predetermined laws* which fix them in their form, and *physicochemical conditions* which cause them to appear. In a word, the vital phenomena is preestablished in its form, not in its appearance. We ought therefore to understand that these phenomena of life can be influenced only via the material conditions which manifest them, but which are not really their cause.

*See my *Rapport sur la Physiologie Générale,* 1867; and *Les Problèmes de la Physiologie Générale* in *La Science Expérimentale.* Paris, 1878.

We need not preoccupy ourselves with final causes, that is to say with the intentional purpose of nature. Nature is intentional in her purposes but blind in execution. We act on the executive side of things and address ourselves to material conditions; it could be said that we act simply to set the stage for nature.

As to the laws, we can learn them; observation reveals them to us, but we are powerless to change them.

Anticipation is made possible by the knowledge of the laws; the observational sciences can not go beyond this.

Action, which belongs to the experimental sciences, is made possible by the determinism of the physicochemical conditions which cause the appearance of the phenomena of life.

In résumé, determinism remains the great principle of the science of physiology. In this regard there is no difference between the sciences of inanimate objects and the sciences of living bodies.

The END

Fig. 1—A. *Stentor polymorphus,* filled with granules of chlorophyll.

a, mouth
a′, nucleus
a″, pedicle

B. Isolated grains of chlorophyll from *Stentor polymorphus.*

b, whole grains
c, c, grains in the process of division
d, d, division into three or four parts.

Fig. 2—A. Plant cell containing chlorophyll.

a′, nucleus of the cell

B. Isolated grains from the plant cell.

b. whole grain
c, grain in the process of division
d, grain almost completely divided

Fig. 3—A. Amoebas having engulfed green particles.

B. Lymphatic corpuscle of *Lumbricus agricola* having engulfed the same green bodies.

Fig. 4—*Zygnema*

A. Zygospore arising from the fusion of the contents of two cells: male (♂) and female (♀).

Fig. 5—*Pandorina morum*

1, isolated zoospore
2, 3, 4, phases of the conjugation of two zoospores
5, oospore

Fig. 6—*Spirogyra*

Passage of the protoplasm from the male cell (♂) to the female cell (♀).

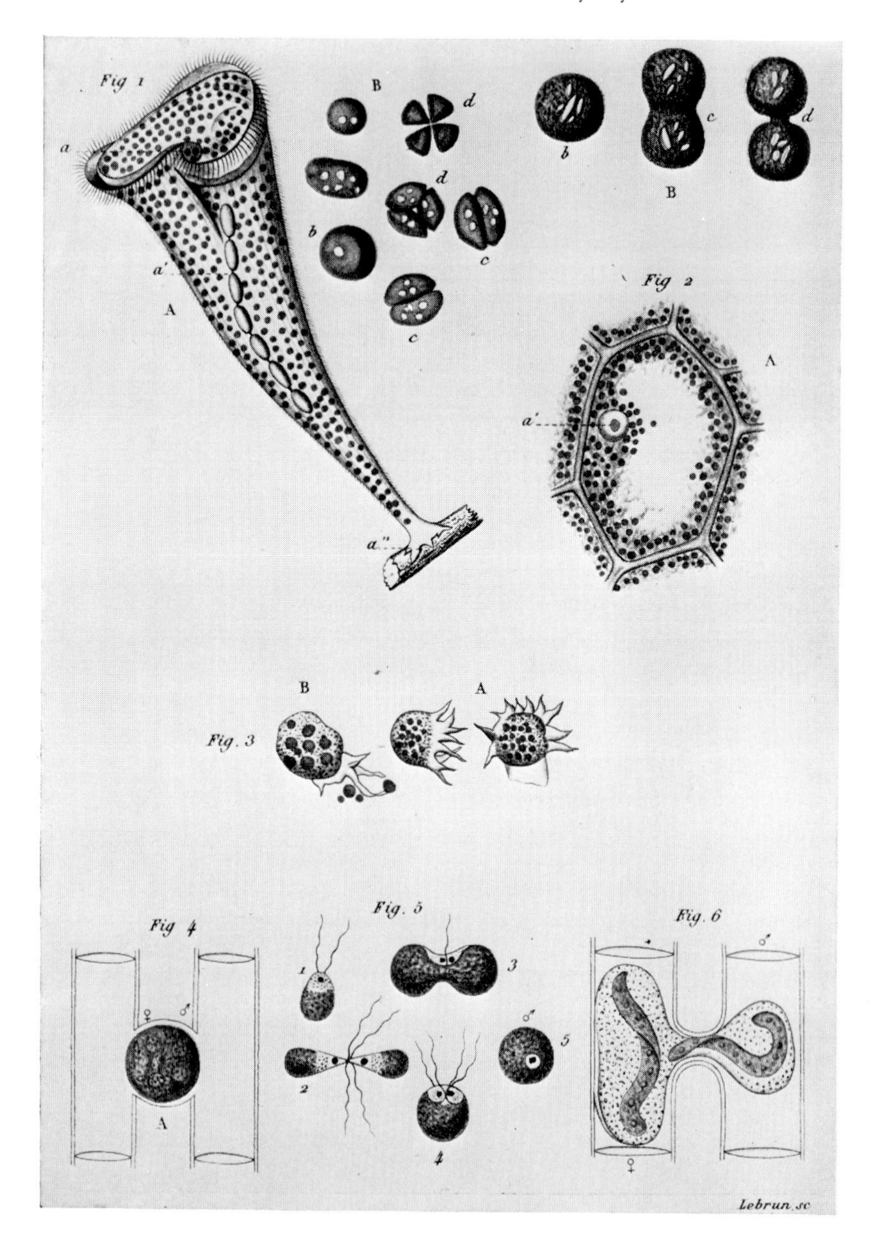
Fig 1
a
a'
a''
A
B
b
c
d
Fig 2
A
a'
Fig. 3
B
A
Fig 4
♀
♂
A
Fig. 5
1
2
3
4
5
♂
Fig. 6
♂
♀
Lebrun sc

APPENDIX

I*

THE CREATION OF LABORATORIES characterizes a new era in which the cultivation of physiology as well as other experimental sciences was introduced.

The installation of these facilities, where there is assembled a more or less complete instrumentation, responds to a double necessity: the requirement of teaching and the requirements of research.

Teaching has its full effectiveness only on condition of showing the objects and phenomena that make up its material. As far as the physical sciences and zoology itself are concerned this condition has been so well appreciated that even in secondary institutions, manipulative exercises have been introduced for the students in so far as possible. Purely *theoretical* or *mental* teaching of the experimental and natural sciences is an absurdity and a relic of ancient scholasticism. What is true for secondary instruction is even more so for higher education, and courses in physiology in particular are nowadays illustrated by experiments and demonstrations which the professor increases to the extent that his program and his resources permit.

The necessity for research laboratories is even more evident, although some persons oriented toward the past bring up, as an argument against our demands, the greatness of the discoveries of our predecessors, and the slenderness of the resources at their disposal. Neither Lavoisier, nor Ampère, nor Magendie had well equipped laboratories. This is true, but these were obstacles over which their genius triumphed, but from which it did not profit. A special installation avoids the loss of time and makes for efficiency in the use of our facilities. It ought to be such that once an experiment is conceived it can be carried out easily and rapidly.

*Opening lecture, P. 1.

Thirty years ago, when we had conceived the idea of an experiment, with what difficulties, with what loss of time did we finally succeed in carrying it out! We experimented in inappropriate places, in an office, in a room, on animals obtained by stealth, or indeed we lost entire days running after our experimental subjects, betaking ourselves to the abbatoirs, to the butchers. It is impossible to transform such a state of things into a model of good scientific administration.

It is necessary for laboratories to place within reach of the experimenter, and at his hands, the subjects and the necessary supply of instruments, so that he will not be deterred by the difficulties of carrying out the research that he has conceived.

II*

DEVELOPMENT NOT TO BE DISTINGUISHED FROM NUTRITION

In the admirable introduction that opens his *Histoire du Règne Animal* Cuvier, induced to speak of the origin of living beings, expresses himself in this way: "The origin of organized beings is the greatest mystery of the organic economy and of all nature."

In reality the mystery of the origin of life is no more obscure than all the other mysteries of life, and it is not any less. From the time Cuvier wrote the preceding lines, many efforts have been made with the design of piercing the clouds that hover over these phenomena. The fruit of so much work has not been, as is thought, to explain the inexplicable, *but only to prove that the phenomena of the origin of life are not of a different kind, not of a more impenetrable obscurity than all the other manifestations of the "organic economy, and of the whole of nature."*

This is already a considerable advance. To relate to the same principle the things considered until then of a different kind, such as the *origin of the beings,* and the *maintenance* of their existence is to have made progress comparable in some degree to that which was realized in another branch of our knowledge the

*Note for page 24.

day Newton proved that weight was a particular case of universal attraction.

A single law, in fact, dominates the manifestations of life that *begins* and the life that *maintains itself;* it is the law of *development.*

Like all ideas whose meaning is derived slowly, the idea of *development* was proposed everywhere and defined nowhere. It acquired its real meaning and application only through the work of the contemporary embryologists. Founded on precise facts, it must henceforth be considered not as one of those banal generalities created by the systematic mind, which are so often in vogue in the sciences, but as the most general conclusion from the discoveries made in the last fifty years.

In order to see how far we have progressed, let us consider the starting point. The phenomenon of the appearance of a new being, engendered or created by whatever mechanism, has always been isolated, separated from all the other vital manifestations and considered as a different and superior kind. Nothing was seen beyond that first moment when the life of an individual was lighted up within the germ. There seemed to be at this point a physiological discontinuity: *"Hic natura facit saltum."*

In reality this hiatus was the only one, and the being, once animated by the spark, continued to live and develop without interruption, following the continuous pathway assigned to it by strict laws.

Thus the living being presented two mysteries, that of birth, and that of the continuation of life, which develops and maintains itself.

This can no longer be maintained today. The principle of development consists precisely in this affirmation, that *nothing is born, nothing is created, everything is a continuation.* Nature does not afford us the spectacle of any creation, it is an eternal continuation.

Before being constituted in the state of a free being, independent and complete, as an individual so to speak, the animal has passed through the state of the *cell-egg* which was itself a living element, an epithelial cell from the maternal organization.

The ladder of its relationship extends infinitely into the past and in this long series there is no discontinuity at all; at no time does a new being intervene, it is always the same life that continues. An immanent propulsive force, reinforced by fertilization, carries the element through all the metamorphoses, through childhood, adolescence, adulthood, decrepitude and death; directing it in this way toward the accomplishment of a plan drawn up in advance. The characteristic of all the phenomena that take place is that they are the result or the consequence of an anterior state, that they are a continuation. This developmental force immanent in the *cell-egg, embodied at its beginning* and communicated to every thing that is derived from it, is the most general intrinsic characteristic of life, and the only thing in it that seems mysterious to us.

Thus what is essential, fundamental, and characteristic of vital activity is this faculty of *development* which so operates that the complete being is contained in its starting point. In this way the necessary unity is established for all vital phenomena, which themselves are the consequence of the developmental force, whether it is nutritive or generative.

The work of the physiologists has had precisely the result of abolishing the barriers separating the egg, the embryo and the adult, and bringing to light in those three states the unity of an organism, taken at three different stages of its course but always subject to the same force and governed by the same law.

But this is not the only effect, and the *principle of development* is not yet adequately characterized by the idea of *continuity*.

So defined, development is not in fact an *actual property* or *a concrete fact;* it expresses simply the law that regulates the succession and the chronological linkage of the vital *facts* that play their role in the organized being.

Is it possible to characterize this law by its means of execution? This is what we shall see.

The law of development applies not only to the whole being, to the individual, but also to each of its parts. It is an *elementary law*. It governs the anatomical element as well as the complete being as a whole, and this is apparent a priori, for there is nothing essential in the whole being which is not present in its component parts. The zoological individual, the animal, is only a federation of elementary beings, each developing on its own account. It was a long time ago (1807) that this idea was expressed by a man who was a thinker as well as a great poet and a wise naturalist; Goethe, meditating on the teaching of Bichat, wrote: "Every living being is not an indivisible unity but a plurality; even when it appears to us in the form of an individual, it is a union of beings, living and existing by themselves."

These elementary organites behave like the individual, their existence is divided into the same periods; they grow, rise, and fall, describing a trajectory of fixed form.

When the attempt was made to fathom what is essential in the life of a being, it was seen that *nutrition* was its most general and most constant characteristic. But nutrition, that is to say the perpetual communication of the anatomical element with the medium surrounding it, this continuous relationship of exchange of liquids (nutrition properly spoken) and of gas (respiration), nutrition, we say, is capable of variations. Growth, the state of full development, and decline, correspond to relative variations in this exchange, in which the *milieu* receives less, as much, or more than it gives to the organism. It is thus impossible to separate the property of nutrition from the conditions in which it is exercised; it is impossible to separate nutrition from growth, from development, and the succession of the ages, that is to say from development. *Development* is the continuous sum total of these variations in nutrition; it is nutrition considered as it exists, embraced at a single glance across time. This development, or law of the variations in nutrition, is from the viewpoint of the philosophers what is most characteristic in life. It is something comparable to the law of movement of this machine that is the living being, and expresses the activity of this being, like the trajectory expresses in mechanics the circumstances of the activity of a moving body. It can thus be imagined that the elementary being as well as the complex being is thus embarked upon a sort of ideal trajectory that imposes its development upon it. The idea of development is the idea of this trajectory, of this law that governs the living being; it is not a fact or a property, it is an idea. The fact and the property are nutrition with its variations, the idea is development, it is the concept of the whole of these successive variations.

The generation or the birth of the being does not make a breach or interruption along this continuous pathway. There is no reason to impose a beginning upon development. Embryological and ovogenic studies have indeed given evidence on this point. The being that is born is not a new creation; at its origin, in the earlier evolution of the beings from which it has arisen, and of which it is the continuation, it has imbibed, by a sort of habit or physiological remembrance, the necessity of the path it must follow. In a word, it is the *same evolution* which lasts and develops.

But in reality the sole fact that can be grasped, that is actual and real, is nutrition. It was a mistake to contest this view and to attempt to separate "nutrition, which simply maintains, from development, which grows, augments and adds."

Contemporary works have had just this result, to confuse "the phenomena of development of what is born with the birth of this object." At the time when Saint Thomas of Acquinas established the division of the *soul* or *vegetative faculty* into three different faculties, *nutritive, augmentative,* and *generative,* he gave proof of a profound philosophical sagacity for his time. As much can be said of Broussais when he distinguished *nutritive irritation* from *formative irritation.* But today, the barriers erected between nutrition, development and generation have fallen under the efforts of those who have followed the earliest phenomena of the appearance of the beings.

It has ben said (page 26) that development characterizes the living beings and distinguishes them absolutely from inanimate objects.

From this there derives a different method in the two kinds of sciences, physicochemical on the one hand and biological on the other. The physicochemical object has an existence in the present, there is nothing beyond its present state; the physicist does not have to concern himself with the beginning or the end. The body manifests all its properties.

On the contrary, the living being, besides what it manifests, contains in the latent state, in potential, all the manifestations of the future. To apprehend it at present, in the very act, is not to apprehend it completely, because it has rightly been said of it that it was a "perpetual future." It is a body on the march; it must be apprehended during its progress and not alone at the stages along its route.

The necessity of this point of view is imposed on natural history in the proper sense of the term. To classify a being it is necessary to have followed it during its whole development; it is not enough, as Cuvier said, to take it at a given time, even at the moment of its most complete development in the adult state. It is not true that the being carries with it, "inscribed for all time within its organization, the characteristics that classify it."

Now we see the necessity for this same point of view in physiology, the study of the phenomena of life as it *develops,* as well as of life that *maintains* itself.*

III†

Examples of the longevity of seeds are quite numerous, but some reservations must be made in the particular case of the so-called *mummy's seed.*

This is what Berthelot says in the *Revue archéologique* for December 1877, p. 397.

> The allegations regarding seed in mummies that has germinated and fructified are today recognized as false by botanists and agriculturists; the individuals who made these trials at other times were dupes of the Arabs and the guides. But no sample ever collected under authentic circumstances has ever germinated.

It is clear that this reservation about the fact of germination of seeds from the Egyptian tombs does not extend to other well-documented examples of storage of seeds and does not modify in any way the conclusions we have drawn therefrom.

---

*This note is the development, as faithfully as possible, of ideas frequently expressed by Claude Bernard in his conversations, which he proposed to reproduce in the appendix. (Dastre)

†See p. 51.

IV‡

The first substance engendered under the influence of life that has been reproduced artificially is *urea.* Wöhler obtained it by keeping a solution of ammonium cyanate at the boiling point for several moments. The transformation of this salt into urea takes place by a simple process of isomerism.

Later it was formed by the reciprocal action of chloroxycarbonic gas and ammonia. This latter reaction established the true constitution of urea, by showing that this substance is the amide of carbonic acid.

Piria next synthesized salicylic hydride (essence of meadowsweet) by the oxidation of salicin.

Later Perkins, causing a mixture of acetyl chloride and sodium acetate to react on this salicylic hydride, effected its conversion into *coumarin,* the crystallizable principle found in Tonka beans.

Piria produced benzoyl hydride (essence of bitter almonds) by the distillation of a mixture of benzoate and formate of lime.

Cahours synthesized a product entirely identical to the oil of *Gaultheria procumbens,* an essence endowed with a very agreeable odor, elaborated by a plant of the family of the heathers that grows in New Jersey; this essence is none other than methyl salicylate.

Salicyclic acid was formed in 1872 by Kolbe, who made carbonic gas act under certain particular temperature conditions on completely dry sodiated phenol (sodium phenate). Dessaignes remade hippuric acid by the reaction of benzoyl chloride on zinc glycocol.

Berthelot achieved the synthesis of formic acid, or rather, of potassium or sodium formate, by the direct union of carbon monoxide and these alkalis. In these circumstances he produced a formate from which formic acid was isolated by treatment with a more stable mineral acid.

Perkins and Duppa on the one hand and Schmitt and Kekulé on the other, have synthesized the malic and tartaric acids found in a great number of fruit acids by the action of potassium on mono and dibromated succinic acids.

---

‡See lecture VI, p. 147.

Up to the present it has not been possible to achieve the direct synthesis of any organic substance from its constituent elements. Up to now it has been possible to produce only indirect syntheses. Thus free carbon and hydrogen combining, as Berthelot demonstrated, under the influence of the electric arc, gives acetylene C^4H^2; this, by fixing hydrogen, engenders ethylene C^4H^4 which, fixing water, gives rise to alcohol. The synthesis of alcohol, an organic product, is thus an example of those indirect syntheses of which we have spoken.

V

FIXATION OF NITROGEN IN ORGANIC COMPOUNDS

By Berthelot*

The experiments of Berthelot† tend to establish that under conditions comparable to the usual atmospheric conditions there can be a fixation of the nitrogen of the air onto ternary organic compounds such as cellulose and starch. Atmospheric electricity acting through differences in potential manifested a short distance from the ground, might make the nitrogen enter into the hydrocarbon materials in the plants. The inference (not yet verified) that these studies permit is that the influence of cosmic agents would be capable of transforming ternary substances into nitrogenated combinations. Such a phenomenon would shed a strong light upon the problem of organic syntheses.

Whatever these long-range inferences may be, here are the exact results of Berthelot's remarkable experiments.

To produce differences in electrical potential maintained within a designated space, Berthelot employed an apparatus composed of two jars of thin glass, the one covering the other so as to leave a space or chamber in which are placed the substances to be studied. The inner jar is covered over its internal surface by tin foil, constituting the positive plate of the condensor, the outer jar is covered on its external face with another sheet of tin foil

*Note relative to page 147.

†*Annales de Chimie et de Physique,* December 1877.

constituting the negative plate. The system rests on a glass plate varnished with lacquer. All is arranged so that the two jars are, in addition, as close as possible to each other.

The external surface of the little jar is covered over its upper half with a sheet of Berzelius paper, weighed in advance and moistened with pure water. The other half of the same surface is coated with a layer of a syrupy solution of dextrin, titrated and weighted, in circumstances that made it possible to know exactly the weight of the dry dextrin employed.

The whole system of jars was sheltered from dust under a glass jar.

Things being arranged in this way, the internal plate of the little jar is connected with the positive pole of a pile formed of five Léclanché cells arranged in series; the external plate of the large jar is connected to the negative pole. The difference of potential between the two plates was thus kept constant. These differences in potential are absolutely comparable to those of atmospheric electricity acting at a short distance from the ground.

Before the experiment nitrogen was determined in the two substances. There was found:

Paper	0.10
Dextrin	0.17

After the experiment had been kept up for seven months, determinations gave:

Paper	0.45
Dextrin	1.92

There is thus a fixation of nitrogen. The space between the two jars, and in consequence the value of the potential has an influence in the phenomenon, for the distance between the two jars being tripled, after seven months, everything being equal, Berthelot found as quantities of nitrogen:

Paper	0.30
Dextrin	1.14

The fixation of nitrogen on immediate principles, cellulose and starch, is thus placed beyond the reach of doubt.

Light is of no value in the phenomenon, everything taking place just the same in absolute darkness.

The attempts of Berthelot to produce chemical reactions different from these, with the same difference of potential have not succeeded.

VI*

The existence of *Bathybius* has been contested, and has given rise, in recent years, to a controversy that is not yet at an end. The naturalists on the second *Challenger* expeditions have considered this matter as a gelatinous precipitate of calcium sulphate; more recent studies contest this opinion.

We need not participate in this quarrel. Besides *Bathybius* there are already enough well-recognized protoplasmic beings for the existence or nonexistence of this one to be able to make any difference in our conclusions.

VII

After the preceding account, is it possible to relate us to a philosophical system? It would be tempting to include us among the materialists or physicchemists. We do not belong to them at all. For, looking at the actual state of things, we accept a *special modality* in the physicochemical phenomena of the organism. Are we among the vitalists? Again, no, because we do not accept any form of executive outside of physicochemical forces. Are we thus empirical experimenters who believe with Magendie that the fact is enough and that experimentation has no need of a doctrine for its own direction? No more so; we find on the contrary that it is necessary, especially today, to have a criterion by which to judge and a doctrine to unite all the facts of science.

What then is this doctrine? *Determinism.* It is illusory to pretend to arrive at the causes of phenomena by the mind or by the matter. Neither mind nor matter are causes. There are no causes for phenomena, and in particular for the phenomena of life, and for all those that have a *development,* the notion of cause dis-

*Note for Page 136 and 214.

appears, since the idea of constant succession does not involve in this case the idea of dependence. The phenomena of development are linked in a rigorous order, and nevertheless we know that the antecedent does not with certainty control the successor. The obscure notion of cause ought to be referred to the origin of things; it has no sense other than that of primary cause or final cause; it ought to give way in science to the notion of relationships or conditions. Determinism fixes the conditions for phenomena, it makes it possible to anticipate their appearance, and to evoke them when they are within our reach. It does not give us an account of nature; it makes us masters of it.

Determinism is thus the sole scientific philosophy that is possible.

In truth it does forbid us the search for the why; but why is illusory. On the other hand, it exempts us from doing like Faust, who after affirmation, plunges into negation. Like those religious who mortify the body by privations, we are reduced, to improve our mind, to mortify it by the deprivation of certain questions and by the admission of our impotence. While thinking, or better, while feeling that there is something beyond our scientific prudence, it is thus necessary to throw ourselves into determinism. If after that we let our minds be sheltered from the winds of the unknown, and within the sublimities of ignorance, we shall at least have distinguished between what is science and what is not.